PRODUCTION AND PLANNING
APPLIED TO BUILDING

PRODUCTION AND PLANNING
applied to building

by

R. J. HOLLINS

Fellow of the Institute of Building.

Formerly Principal Examiner in Management Subjects for the Institute of Building.

Formerly Visiting lecturer at College of Advanced Technology, Birmingham and Lanchester College of Technology, Coventry.

Former Chief Planning Engineer—C. Bryant & Son Ltd., Birmingham.

Deputy Managing Director of The Greaves Organisation Ltd.
J. Greaves & Co. (Services) Ltd.
Rotheroe Developments Ltd.
Rotheroe Developments (Coventry) Ltd.
Doric Plastics Ltd.
Cadgate Properties Ltd.
Colinwave Ltd.
Buckhithe Properties Ltd.
Timmeroe Developments Ltd.
West Midland Financial & Development Co. Ltd.
Great Barr Developments Limited.

PUBLISHED BY GEORGE GODWIN LTD., 4 CATHERINE STREET,
ALDWYCH, W.C.2

(A member of The Builder Group)

FIRST PUBLISHED 1962
SECOND IMPRESSION APRIL 1967
REVISED EDITION JANUARY 1971
© THE BUILDER

SBN 7121 4606 7

A/658.99

Printed in Great Britain by
Page Bros. (Norwich) Ltd., Norwich

FOREWORD

The development and future productivity of the building industry are dependent upon the gainful employment of its resources. Such resources include the wealth of manpower, both mental and manual, the availability of materials, the accumulation of capital and the skill and ability to plan, organise and administer the many enterprises that form the complex which is concerned with fulfilling the community's building requirements. This last factor, which is now generally referred to as Management, is undoubtedly by far the most critical in the economic development of the industry. With the growing awareness of more and more individuals of the need to improve the quality and capacity of building management, the provision of suitable textbooks has become a matter of cardinal importance.

As Mr. Hollins reminds his readers in his Introduction, the qualifying examinations for corporate membership of the Institute of Builders now include compulsory papers in Management Subjects and there can be no doubt that this recent and enlightened decision of the Institute will have far-reaching effects upon the future pattern and productive capacity of the building industry. The author has made a valuable contribution to the bibliography of the building industry, for, among existing textbooks concerned with the conduct of building work, few deal with the practical application of management principles to the exacting process of marshalling and gainfully directing the resources of manpower, materials and machinery to economic and specific ends.

The problem facing the industry and associated professions of training a sufficient supply of qualified executives with high-grade managerial, industrial and commercial knowledge will inevitably take time to resolve. However, much can be done if the underlying principles are recognised, studied and applied to the day to day conduct of individual enterprises and projects.

Mr. Hollins devotes the bulk of his work to the concept of Planning and the application of planning principles to the process and prosecution of building work. It would be surprising if his admirable book failed to stimulate the thinking and increase the managerial skill and capacity of those able to absorb his distillation of practical management expertise.

D. E. WOODBINE-PARISH, C.B.E., F.I.O.B.

PREFACE TO FIRST EDITION

To aid the ambitious executive in the building industry, many Colleges of Technology now offer courses in Management and Planning.

The Institute of Builders has recently introduced its new examination structure and syllabus. Management subjects proper form Part II of the final examination and no provision is made for exemption. In deciding the scope of this book, I have in general, limited it to the current syllabus for the Management Exercise paper of the examination.

The book is also intended to be an aid to senior graduate students and executives of medium-size building firms, who have to formulate policies and procedures in order to improve the efficiency of their organisation. It is my hope that this publication will give some help to those who believe that a higher standard of living can only be achieved from higher productivity and to those who are prepared to do their part in the organisations of building contractors of this country.

My reason for lecturing and writing on this subject is really simple; it is the firm belief that much can and should be done on this question of productivity in the building industry. I am convinced there is only one way to do it:

PLANNING IN ADVANCE

Building has always been a process demanding co-operation between client, architect and contractor. The success of building operations depends upon perfect understanding between the members of the team. This was never more true than it is today, when problems of economics, materials and labour provide such hazards in themselves that there is no margin for inefficiency within the team.

It is still commonly assumed that the sooner building operations are begun, the sooner they will be finished. The opposite is nearer the truth, for the longer the time spent, within reason, in planning and preparation, the more speedily will the work be completed.

In the past, insufficient attention has been paid to training building industry management staff to ensure that they are effective leaders, as well as proficient in their knowledge of the trade. In securing their jobs, they may either have worked their way up or have been dropped in from the top. In neither case has proper importance been given to the possession of the mental outlook which their position demands.

Management today calls for a higher standard of qualifications and ability than are likely to result from length of service alone.

Those who are preparing to devote their career to the practice of modern management techniques will, I hope, acquire useful knowledge from the books recommended in the Bibliography, then use their enlarged abilities to qualify for further advancement. They can rest assured that the supply of adequately qualified persons falls short of the demand.

I express my sincere thanks and appreciation to the Board of Directors of C. Bryant & Son Ltd., for training me in planning techniques and permission to reproduce in this text the management and production philosophy of the company.

I am deeply indebted to The Advisory Service for The Building Industry who together with Urwick, Orr & Partners Ltd., Management Consultants, provided me with the inspiration to write this book.

PREFACE TO SECOND EDITION

In the eight years since *Production and Planning Applied to Building* first appeared many changes have taken place in the building industry, including the introduction of techniques such as Critical Path Analysis, Computerisation, and now Decimalisation. The change to Decimalisation necessitated frequent revisions in the text where imperial measurements were formerly used.

We gave consideration to the introduction of new chapters on Critical Path Analysis and Computerisation but it was felt that the purpose of the book was principally to assist students of management subjects in acquiring a knowledge of the basic principles. Other, more advanced, well-known publications are available as the means of developing further advanced knowledge in these techniques.

One of the things that I personally have learnt over the last few years, in running a building company, is that the understanding of the theory of management is one thing and its practical application is another. A building firm is only as good as the people who work in it and it has become increasingly more important that not only senior executives and managers should be trained to organise, plan and manage their affairs but it is also vitally important that all levels of responsibility throughout the company receive adequate training. In the 1970s no business venture can succeed if its affairs are controlled by 'enthusiastic incompetents'. Excellent opportunities and facilities are now provided by the Construction Industry Training Board and I give it to you in strong terms of recommendation to consider seriously introducing a comprehensive executive development training programme at all levels.

The reception that has been accorded this book by students and executives of building firms everywhere has been deeply gratifying to me and it is my sincere wish that the book, as revised, will continue to be of service to these old friends, as well as to yet another generation of students in modern management techniques.

I hope that this new edition of *Production and Planning Applied to Building* will serve as an aid to all students and executives of medium-sized building firms in the difficult years that lie ahead—years that may well prove decisive in shaping the future of the building industry of Great Britain.

I am deeply indebted to Mr. W. T. Patrick, FIOB, ABldgSI, FIBICC, Acting Head of the Department of Building at Hall Green Technical College in Birmingham and Mr. D. Helm, BSc, AIOB, who have been of considerable assistance to me in modifying the text of this revised edition.

R. J. HOLLINS

January 1971

CONTENTS

ILLUSTRATIONS

11

CHAPTER I

INTRODUCTION

... Planning is impossible unless every component of the organisation contributes its share to the total result ... The executive may not consciously use the term, but all his work, whether he knows it or not, is in some way concerned with the unfolding of a plan ...[1]

Each year over £3000 million is invested in Building, this represents about half the annual capital investment of the Nation.

A mere one per cent saving in the costs of the annual building programme would save at least £30 million—enough money to build another 12 000 new houses or another Motorway.

This is a saving that the Building Industry could make comparatively quickly and without much difficulty, providing that it is borne in mind that building operations have become more complex and exacting over recent years.

This has been brought about by the introduction of new materials and methods, in addition to the ever increasing size of large buildings and fundamental changes in structural design.

Not only have buildings changed in design, but also in the management skills and the methods used to construct them. The management skill required to run successfully a nineteenth century building firm is insignificant compared with the skill required to manage the complex organisation of a modern contractor capable of dealing with the fierce business competition today. The great size of many of our larger national contractors today would not have been possible under the management methods of one hundred years ago.

To be concerned with change is to be concerned with the

[1] Marshall Dimock. *The Executive in Action*. New York, Harper & Brothers 1945. p. 124.

future. In management we call this consideration for the future, planning.

Planning is, therefore, a basic executive responsibility.

Planning is an intellectual process introducing creative thinking.

PLANNING DEFINED

Planning is the process by which executives anticipate the probable effects of events that may change the activities and objectives of their business.

By planning they attempt to influence and control the nature and direction of the change and to determine what actions are required to bring about desired results.

Planning is the conscious determination of courses of action, the basing of decisions on purpose, facts and considered estimates.

Without effective plans, action is likely to become merely random activity, producing nothing but chaos.

THE PURPOSE OF PLANNING

Planning building work and controlling building work are so closely linked as to be inseparable. They may even be referred to as the 'Siamese twins' of building management.

Certainly no contracts manager can properly control a project that has not been planned, for the very meaning of control is keeping site operations on course by correcting deviations from the programme.

Any attempt to control without planning would be meaningless.

The old method of discovering errors, mistakes or omissions after the damage has been done and then applying pressure is grossly inefficient.

Through planning it is possible to prevent problems arising or to minimise them, as in the case of a systematic plan for replacing builder's mechanical plant at regular intervals to assure a continuous availability of efficient equipment on site.

In more complex situations, a builder frequently finds that the pressures of problem situations call for extra planning effort. An example is provided by a building contractor who, in 1953, began to face the problem of inadequate production facilities in his Joiners' Shop. The business had grown so that its production requirements were beyond the capacity of its present shop. The top executives therefore planned a series of actions calculated to solve the problem. They eventually set the goal of building a new Joiners' Shop. When the shop was built in 1955, the executives were ready with plans for moving into it from the old location.

Suppose, however, that the firm had not been alert to the significance of the growing pressure on the Joiners' Shop. The result would have been a much more sudden awareness of the need for expansion, due to a hold-up of supply of joinery material on site. There would then have been too short a time in which to take corrective action. Fortunately, the executives concerned had the necessary foresight to look sufficiently into the future. By planning over an adequate time they accomplished the needed changes on an organised basis and without disturbances.

This reminds us that an important ingredient of successful planning is attention to future developments.

Planning thus involves the appraisal and measurement of current conditions and their comparison with those desired or expected.

The modern conception of control is based on sound principles of practical planning and foresight, with adequate means of measuring the actual progress of work against predetermined output standards.

THE TIME FACTOR

Since change is a fundamental part of our definition of planning it follows that time also is a central element, for change occurs through particular spans of time.

Time spans with which planning is concerned range from relatively long to relatively short terms.

Short-term planning is that which is concerned with planning for the comparatively near future. Plans for the next month, or even the next six months are short-term plans.

'. . . Time is the central element in planning and the nature of the planning depends upon the time span considered.

The executive must integrate both long- and short-term

Figure I.

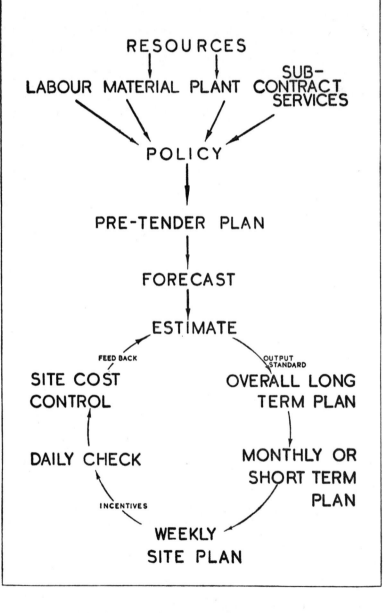

THE PLANNING CYCLE
IN THE
BUILDING INDUSTRY

RESOURCES

LABOUR MATERIAL PLANT SUB-
CONTRACT
SERVICES

POLICY

PRE-TENDER PLAN

FORECAST

ESTIMATE

FEED BACK   OUTPUT
STANDARD

SITE COST   OVERALL LONG
CONTROL   TERM PLAN

DAILY CHECK   MONTHLY OR
                             SHORT TERM
                             PLAN
INCENTIVES

WEEKLY
SITE PLAN

planning so that the continuous stream of effort will result in the harmonious attainment of objectives. . . .'[1]

It is important to grasp that a long- and short-term plan are two aspects of the same continuous process. Success in planning depends on the ability of the executives to achieve an integration of the two.

Short term planning can be successful only if carried out in a context of adequate long-range planning—if the time factor of the long-range plan has been taken into account in the preparation of the short-term plans. For example, the preparation of a plan for the construction of a typical floor of a multi-storey building is in itself a short-term plan, but it is also an essential and closely integrated part of the Master Plan for the construction of the complete building.

OBSTACLES TO EFFECTIVE PLANNING

We are all accustomed to planning, if only with respect to a future social event or an annual holiday. However, even the most complex planning by the individual is likely to be far simpler than that required by even a small business, to say nothing of an organisation with branches all over the world.

External forces such as laws, government regulations, trade associations and the action of competitors may impede or influence the effectiveness of planning.

Managers often do not plan as well or as thoroughly as maximum effectiveness in their work requires. This is not entirely their fault since many factors which lead to incomplete planning are beyond their control. However, managers can improve their planning, first by understanding the obstacles, and secondly by taking specific steps to minimise their influence.

In studying the obstacles to planning in the building industry we shall consider the influence of:

1. The time span covered.

2. Unforeseen or unpredictable events.

3. Mental ability.

[1] Dalton E. McFarland. *Management Practices and Principles*. New York. The MacMillan Co. 1958. p. 72.

B

4. Lack of information.

5. The human element.

6. Costs of planning.

1. *The Time Span*

As the time span of the work concerned increases, the accuracy of planning tends to decrease. This arises in part from the fact that a greater span of time increases the probability of some unexpected event disturbing the plans that have been made. The more remote the future time the manager is considering, the less certain becomes his ability to anticipate every possibility.

A further factor is that the present has a much stronger influence on an executive's thoughts than he may realise, even though he is capable of effective long-term planning.

Present conditions weigh heavily in planning and by overshadowing future needs, may result in errors of judgement.

For example, a contractor encountered piling problems at the start of a job which led to the arrangements for delivery of structural steel being neglected. At a later date it was not possible to keep the site supplied with steel. This resulted in extended delays in the contract despite the satisfactory completion in the early stages of the difficult piling foundations.

The relative difficulty of long-term planning often leads to serious shortcomings in the work of many managers in the building industry. The natural tendency is to concentrate on short-term planning where it is easier to feel more confident of being able to cope with the situation and to leave future problems to look after themselves. It would be wiser to let subordinates concentrate on setting out the work of the moment so that senior management men can devote intensive planning to material and plant requirements of the immediate future. We must remember that delegation of responsibility is the hallmark of a successful leader.

2. *Unforeseen and unpredictable events*

In building, as in all human affairs, it is not possible to foresee all that may have serious consequences for our plans. Therefore, planning must be based on what is known in the present and modified (if necessary) by what one can judge will develop in the future.

Men vary in their ability to see ahead and foresight is often listed as one of the desirable qualities of the successful executive. An extreme example of an unexpected event can be found in the case of a firm of industrial engineers, two of whose directors attended a conference in the north of England and travelled back together by car. They became involved in a road accident and both were killed. The situation in the company which thus lost two top executives was critical. No thought had been given to the need for a proper programme of executive development. Without a proper training programme the organisation had no immediately promotable senior executives. No one had been trained from the ranks of middle management to take over the responsibilities of the board. Because of this lack of foresight and planning the company was forced to advertise in the hope of being able to fill the vacancies at once from outside.

It was not surprising that the new directors so hastily appointed were not adequate replacements for the two who were lost and the effect was apparent throughout the organisation for several years afterwards.

3. *Mental Ability*

Urwick has said that '. . . planning is fundamentally a mental disposition, an intellectual process. . . .'

Planning requires a difficult kind of thought process—the ability to arrange a complex array of ideas, and to see various possible combinations of effects. Thought requires effort *and not all executives enjoy it*. It often involves the painful contemplation of unfortunate and undesirable events and arouses reflections on errors of the past. To those who are inclined to be optimistic in their outlook, planning requires that pessimistic occurrences also be anticipated. The executive who by habit is motivated by action is not favourably inclined to sit at his desk and think.

Creative ability is always in short supply. It is usually best drawn forth by a comfortable and relaxed atmosphere, although in our industry, strains, pressures and tensions may make it difficult to find one. Intellectual activity is frequently ridiculed by those who lack respect for it and there are cases well known to the writer of executives who are so sensitive on this point that they go to great lengths to avoid being caught thinking!

4. *Lack of Information*

Planning may also be affected by lack of sufficient information, or by deficiencies in its accuracy and quality. The quantity of information required, as well as its quality, is partly dependent upon the time period involved. There are always limits to the information that can be obtained. There are also limits to the minimum amount of information required to produce a realistic plan. A balance has to be struck between acting on totally inadequate information, or not acting at all because the builder is waiting for complete information.

Experience teaches the executive how to judge when he has enough information to form a reasonably practical plan, and how to know when a plan is unreliable by reason of lack of information that has not yet been obtained.

No builder should be afraid to ask for information, and, more important, neither should he be afraid to refuse to start work until he has all he needs. The Client and the Architect will have long forgotten the commencement date of the work by the time the opening ceremony arrives. If the builder expects to have the building ready for the opening ceremony he has got to have the information beforehand and an opportunity to think.

5. *The Human Element*

One must always bear in mind that planning is not a cure-all, nor does it deprive site supervision and foremen of any of their authority. In no degree does it minimise the foreman's position, but on the contrary should *strengthen* it. Planning makes it possible to utilize fully the existing capacity of the company. However, it should always be remembered that a system can be no better than the personnel operating it.

6. *Costs of Planning*

Planning requires money as well as time. Planning costs include not only the salaries of executives who plan, but the costs of controlling and maintaining the plans, and the costs of false starts that result when planning is incomplete or incorrectly done—as it often will be. Planning costs are overhead costs; that is they are indirectly related to the quantity of work being undertaken.

It should be appreciated that planning is not a 'frill' or luxury to be enjoyed by large companies who can absorb the costs; on the contrary, it is a prime necessity of any company which wishes to operate efficiently. The fact that many companies have in the past operated without good planning techniques, and do still, proves nothing. The end result would be improved by its use, if they did plan.

Planning should never be anything more than a means to an end—one of the most important tools of economical construction. If it fails to be this, it should be disregarded—quickly. In other words, the search for better ways of doing things is never ending. Today's methods, however successful, can never be taken as wholly right. They represent simply the best efforts of the moment. Tomorrow will bring improvements in these methods.

MAKING PLANNING EFFECTIVE

The problems of planning can be handled by the planning process itself. In other words, planning must be planned for. The Mayor of one large city in America decreed that every official on the city payroll had to set aside the last half-hour of his work day for planning. All work had to be set aside, and the time devoted to thoughtful contemplation of goals, problems and future constructive activity.

Although it is not usually possible for everyone to be creative at the same time each day, or day after day, it is important that planning should be incorporated into the work pattern of executives as a matter of habit.

Each builder must work out for himself the methods of planning which work best for him. It is possible to suggest planning procedures, similar to those which some builders already employ and which give promise of being generally useful.

Some of the steps of analytical planning are as follows:
1. Define the problem to be solved.

2. Get as many relevant facts as possible; organise the information available.
3. Analyse the information.
4. Select the alternative courses of action.
5. Weigh the alternatives and decide on which one seems best to meet the problem.
6. Put the decision on record.

CHARACTERISTICS OF EFFECTIVE PLANNING

Planning enables a company to be competitive with others in the same industry. Planning may involve expansion of service departments such as mechanical plant maintenance; changes in methods, for example standardisation of formwork design in reinforced concrete work; and so on. Improvements in methods of construction are the keynote of progressive management in building.

According to Urwick the characteristics of a good plan are:
1. That it is based on a clearly defined objective.
2. It is simple.
3. It establishes standards
4. It is flexible.
5. It is balanced.
6. It uses available resources to the utmost.

Planning is an executive function which helps to provide purpose and direction for the members of an organisation. In this chapter we have considered the major themes of planning. The following chapters on all aspects of this executive function, from policy planning at board level down to routine site planning, will elaborate on the planning concepts we have considered.

CHAPTER II

POLICY PLANNING

'The formulation of policies . . . as one of the most critical tasks . . . is so pervasive and far reaching in its effect upon the affairs of a going concern that it challenges the finest qualities of managerial talent. . . .'[1]

We have already considered that one of the most important current challenges to a modern builder's business comes from the ever-changing demands made upon it.

The old-time management with its limited foresight and rule of thumb methods cannot cope with the complexity of modern building today. By force of circumstances, we have to keep up with industrial progress, research, atomic development, and a whole host of current political, economical and social conditions.

From the building trade viewpoint and in a practical sense, the long-term plan may be out of reach and idealistic, but theoretically, in accordance with good business practice, it is a goal to be accomplished. Any enlightened business management whether in the building trade or not, gives consideration not only to the prime object of making profit, but to the satisfaction of those personal objectives of employees which are legitimate and should be encouraged.

It cannot be too strongly emphasised that good personnel morale is obtained when both employee and employer jointly associate themselves with the common objective of the business, from which benefits are, or should be, shared by all. Before these fundamental principles of modern management can be applied to any company in our industry, the objective or purpose of the builder's business has to be properly laid down. This object or purpose is called 'The Policy'.

[1] E. Peterson & E. G. Plowman. *Business Organisation and Management.* R. D. Irwin Inc, 1953, p. 371

POLICY DEFINED

A Policy is a guiding 'principle' of intent. The most important characteristics of a good policy are that it is:

1. Based upon a principle which has been established after due consideration of all the factors involved.
2. Relevant to the company's current activities as forming a guide to executive action.
3. Carries the authority of the Board.

Policy is not something that is immune from criticism, something that needs no justification, or to be looked upon as though it possessed the character of natural law. In point of fact, changing conditions call for changes in policy. More businesses fail by hanging on too long to what were once, but are not longer, good policies than through formulating new ones. Careful policy planning keeps an enterprise alerted to change by anticipating the need for adjustment. When changes occur, attempts must be made to set out afresh the policies of the company with precision to ensure that they are based on a comprehensive, logical and self-consistent set of principles for the conduct of the company.

There can be little doubt that if some of the companies which have gone into liquidation over the last ten years had given more thought to their policies they would probably be in existence today. It may be found on occasion that for reasons beyond control it has suddenly become necessary to modify policy very substantially. Government action, for example, may result in a great reduction in certain types of building work. In order to keep the labour force together and to avoid a heavy trading loss, the directors may have to formulate a policy to undertake other types of work, or to enter new areas. In order to be effective, such changes in policy may have to be made very rapidly. An able and energetic Board will be able to cope with this kind of situation. If, however, it is not satisfactorily dealt with, serious deterioration in business is then inevitable.

POLICY AND MANAGEMENT COMPARED

It is the task of Policy to lay down strategy to indicate objectives, establish priorities, and set the time scale. It is also part of policy to lay down the broad limits in money, manpower, and resources generally within which the programme of the company must be carried out. In comparison, it is the part of management to create conditions which will bring about the optimum use of all resources available to the organisation in men, methods, plant, materials and specialist services etc.

GENERAL TRADING POLICY

It is most important that careful consideration should be given to the type of work to be carried out by the builder in the future. It is far safer, apart from being more scientific and businesslike, to establish clearly from past results the best and most profitable kind of construction suited for the firm. As a result of this, a proper tendering policy can be established to guide the decisions of the Board of directors when considering invitations to tender. This is not as difficult as may appear at first view. The Contractor can carefully analyse past records, abstracted under different headings of completed work such as schools, houses, factories, offices and so on. The financial results, whether profit or loss, should be stated for each contract. Gradually as the information is compiled, a definite pattern will take shape and the most profitable type of work can then readily be seen. This system of analysis can also be carried out under the following sub-divisions:

(a) Architects.
(b) Local Authorities.
(c) Geographical Locations.

The pattern will then show ideal business contacts and worthwhile areas of operating and so on. The experience of the staff may affect the type of work that the company will be engaged upon in the future. Here again, proper personnel records showing the experience of each employee will assist greatly in reaching major decisions of this nature.

Accurate knowledge of the geographical complications, communications etc. involved in carrying out work in various areas should be investigated.

The meteorological authorities will provide information relating to average rainfall statistics which indicate the amount of inclement weather that can be anticipated throughout the country.

The amount of 'wet time' may vary considerably in different localities and therefore a contractor would be well advised to carry out some form of research into this when considering taking on new work in new areas.

The local availability of labour in areas influenced by the upsurge in large industrial developments could be another major reason for deciding not to undertake certain work.

FINANCIAL POLICIES

Financial planning in a business demands not only research to forecast future economical activity but research to determine the amount of funds required and the most profitable method of adequately financing the undertakings during a future programme of expansion.

Financing of development and expansion out of undistributed earnings seems now to be almost a thing of the past. Income Tax and Profits Tax absorb so much of a company's earnings that there is little left for further development.

The following illustration is quite hypothetical but it will make the principle clear:

Supposing the Contracting Company has:	£	£
Issued Capital		2 000 000
Turnover		4 000 000
Surplus after statutory provision for depreciation		280 000
Of this surplus, Income Tax and Profits Tax will absorb approx.	180 000	
The Shareholders may reasonably expect 5% less tax on their investments i.e.	55 000	
		235 000
Leaving a balance of		£45 000

The directors would on the above showing have at their disposal the sum of £45 000. This would have to provide for any special reserves for depreciation (this being necessary because the replacement cost of plant and equipment and buildings is so much higher than the 'book values' on which the statutory tax allowances are based), and for any other reserves that may be necessary and for development. Clearly, if the Builder has only $2\frac{1}{4}\%$ of his issued capital available for all these purposes, the expansion which he can finance 'out of his own pocket' is negligible.

Any development will have to be financed by fresh capital and it is difficult to see how this capital is to be obtained. In the long run, capital comes only from savings. It is clear that companies, particularly in our industry, cannot save. This is a subject that is, undoubtedly of great concern to the directors of very many small, medium and large businesses today.

Clearly, any future programme for expansion must be preceded by systematic consideration of the following:

1. In x years' time what will be our budgeted turnover?
2. What increase, if any, in working capital will be required?
3. What will be the approximate amount of capital invested in plant?
4. What will the $\%$ overheads be?
5. How are we going to raise the additional capital and how much can we afford to pay in loans to secure it?

THE ORGANISATION STRUCTURE

Relatively few changes in major policy can be made without some adaptation of the structural organisation of the company.

An organisation chart or 'family tree' of the company must be prepared showing clearly the lines of authority and communication. Progress and development will from time to time necessitate alterations in the organisation. It is of the utmost importance for the efficient and profitable running of the concern that those alterations are anticipated well in advance. The builder who plans to expand must ask himself:

1. Will the organisation be suitable to undertake this work.
2. What alterations will be needed?
3. Who needs additional training and when?
4. What additional staff do I need?
5. Shall I recruit from outside or can I start training from within the Company?

TRAINING AND EXECUTIVE DEVELOPMENT

An important policy decision has to be made when continuous development and expansion creates vacancies in the management structure. The problem of selecting a suitable candidate from the company's staff and planning a systematic programme of training in order to develop the executive into a fit and proper person to undertake these additional responsibilities has to be considered.

Many builders subscribe to the 'sink or swim' philosophy, presumably under the mistaken belief that the only way to learn is by experience and 'the deep end' is always the place to start!

There are many cases where young, energetic men have been gradually suppressed by being required to assume duties and responsibilities prematurely and without having received proper training. In consequence, by general stages of frustration and resignation, they become totally unfit for further senior responsibility.

A systematic appraisal of the personnel in each department, in order to prepare a short- and long-term training programme, will ensure that one potentially promotable executive has received adequate training for each possible vacancy in the line of succession in the 'family tree'.

SALES, MARKETING AND PUBLIC RELATIONS

In the building industry, systematic and planned public relations is still in its infancy even in the larger companies and virtually non-existent among the medium and smaller concerns.

Salesmanship, Marketing and the techniques of good public relations are subjects far too involved and specialised to form part of this book. Nevertheless, the policy of the builder must include for some form of systematic advertising in trade journals, local or National periodicals and so on. The cost of advertising can be high but a figure should be set aside annually to cover this important aspect of good business practice.

'Out of sight, out of mind' is a very true saying. With today's keen competition no builder can afford to keep his name out of the public eye for very long.

RESEARCH AND DEVELOPMENT

The development of equipment used in our industry has been comparatively slow compared with the rate at which new equipment is developed in engineering industries.

Building plant manufacturers complain that they get insufficient constructive criticism of their equipment. In order that improvements may continually be made with our mechanical and non-mechanical plant, some form of planned research should take place within each building company. It should be the policy of the builder to investigate, adopt, adapt and improve his equipment. This may be done by a 'Suggestion Box' and thereby provide a continuous means of communication for ideas of improvement.

In the larger organisations, proper Method Study techniques can be applied and the use of trained Work Study Engineers in this direction can result in enormous savings.

THE BOARD OF DIRECTORS' POLICY STATEMENT

In order that the policies of the board of directors are fully understood and properly communicated to all concerned, some form of Policy Statement can be prepared to instruct and inform the staff of the aims and principles of the company.

As an illustration, an extract from a set of policies for a purely hypothetical building company is given as Figure 2. These have been based on the actual policies of a number of companies. The information is by no means full and comprehensive, but it is hoped that they will be a guide to the nature and scope of what has been in operation in certain wellknown and successful businesses.

Figure 2.

AN EXTRACT FROM A TYPICAL POLICY STATEMENT OF A BUILDING COMPANY

The intention of this example is, by setting out a typical set of policies for a building company, to indicate the content and significance of the policies more fully than is possible in the text. It is hoped in this way to illustrate that policies are statements of the principles in accordance with which an enterprise is to be conducted. The policies here stated will obviously not be applicable directly, or as a whole, to any particular business. They may, however, serve to facilitate the preparation and definition of the policies of companies whose directors intend to lay down as precisely as possible a statement of their wishes and intentions.

(1) GENERAL.

It is the policy of the company to:
(a) Conduct its affairs as laid down in its Articles of Association, with due regard to legal requirements, and in accordance with the highest moral standards.
(b) Take all steps necessary to ensure that the business shall continue active and solvent.
(c) Develop the business in the directions and to the extent determined from time to time by the Board.
(d) Make the fullest possible use of its capital, plant and labour by the adoption of all established techniques and scientific methods available for these purposes in order to increase the real wealth of the Company.
(e) Review continuously the proportions in which the wealth created is divided between its members, its employees and to make such adjustments in these proportions as the Board may judge proper in current circumstances.
(f) Advertise in the national press and by other means as the management may decide; the total cost of advertising and sales promotion to be within limits specified annually by the Board.
(g) Maintain its buildings, plant and other fixed assets in a sound and effective state.
(h) Acquire subsidiary businesses associated with the building industry and manufacturing rights owned by others, and enter into manufacturing agreements with others, as the Board may judge desirable.
(i) Control its operations by means of budgets of income and expenditure of all types prepared each six months on a basis of standard costs.
(j) Review its policies continuously in the light of propositions submitted by the management.

(2) FINANCIAL.

It is the policy of the company to:
(a) Operate within its own resources, avoiding short-term borrowing.
(b) Increase its capital as may be necessary to finance development, and to maintain the ordinary share issue at not less than 70 per cent of the total issued capital.
(c) Depreciate its plant and other assets at a prudently high rate; to maintain reserves of such kinds and at such levels as the Board may determine.

(d) Distribute by way of dividends not more than 55 per cent of the net profit arising from any year's operations.

(3) CONSTRUCTIONAL.

It is the policy of the company to:
(a) To tender for work in areas laid down by the Board as may be judged most suitable in the current circumstances.
(b) Carry out proper and systematic planning techniques at the tender stage in order to submit competitive and realistic quotations for work.
(c) Change in construction methods as may be necessary to benefit from results of research or new techniques.
(d) Carry out work on a fixed price basis as may be considered prudent by the Board.
(e) Minimise the amount of work sub-contracted and employ and train skilled craftsmen to carry out such work as the Board considers necessary.

(4) RESEARCH AND DEVELOPMENT.

It is the policy of the company to:
(a) Devote annually a sum equal to between 8 per cent and 11 per cent of its expected gross operating surplus to research and development.
(b) Conduct its research on the basis of programmes to be reviewed by the Board not less than once per year.
(c) Pursue its research and development in directions where the company's existing knowledge and arrangements for building can be used.
(d) Make every effort to improve existing methods and processes by means of research.
(e) Use, as may be desirable, university, trade and other agencies for carrying out and advising on research.

(5) PERSONNEL

It is the policy of the Company to:
(a) Seek to provide a way of life that will satisfy all its employees at all levels.
(b) Allocate approximately one half per cent of its expected gross operating revenue to Personnel and Welfare activities.
(c) Pay its employees in accordance with any statutory requirements or trade agreements, having regard to their effectiveness, in conformity with plans approved by the Board for the various kinds of work and the responsibility involved.

31

(d) Adopt and apply the principles of Joint Consultation.

(e) Recruit young men from trade schools, technical institutes and universitites, and provide all reasonable facilities for their training and development.

(f) Staff vacant appointments from among the existing employees whenever possible.

(g) Discharge from its service any employee found guilty of gross disloyalty or improper conduct, or of an indictable offence committed against the company.

(h) Allow future growth in labour demand, rather than immediate discharge to correct redundancy arising from technical development.

(i) Discharge only the latest employees, according to as liberal a method as possible, if redundancy is occasioned by economic factors beyond the company's control.

CHAPTER III

PRE-TENDER PLANNING

'Nowadays, people know the price of everything and the value of nothing.' *Oscar Wilde.*
'Remember, in any business you cannot spend prestige.' *H. Plotnek.*

Most Conditions of Contract require the contractor to visit the site before tendering for the work; to satisfy himself 'with the nature of the work' and stipulate that he must include in his tender price for provisions to cover risks and difficulties which may be encountered during the course of construction.

This investigation at the estimating stage is vitally important, because upon its accuracy may depend the securing of the job against competitors, and it is hoped, the eventual realisation of the anticipated profit. An inaccurate opinion based upon vague research into the problem at this stage may well mean losing the job, or even worse, obtaining the job and then a loss instead of a profit in the Final Account.[1]

PRE-TENDER PLANNING DEFINED

Pre-tender Planning is a systematic approach to the problem of anticipating and forecasting the probable costs of construction and to accurately determine these costs at the estimating stage which are not always readily apparent from the Bills of Quantities and detailed drawings.

AIMS AND PURPOSE

The object of pre-tender planning is to reduce, or eliminate, the risk of inaccurate opinions influencing decisions which affect the pricing of the estimate. By harnessing the resources and experiences of the company to guide the Estimator when he is considering items of risk, or more difficult structural

[1] John C. Maxwell-Cooke. *Civil Engineering Contracts Organisation.* Cleaver-Hume Press Ltd. p. 46

C

problems, he is better enabled to calculate a more realistic and accurate assessment for the work involved.

It is of paramount importance that team spirit is created and fostered throughout the preparation of an estimate. Decisions which will affect the cost of the job should be made by the Estimating Department but always in close co-operation with the other members of the builder's construction staff.

PRE-TENDER PLANNING MEETINGS

Informal meetings should be arranged from time to time in order that the more important matters can be discussed fully, and that the facilities of other specialist departments within the builder's organisation can be called upon.

THE CONSTRUCTION STAFF

It is a sound idea for the Builder to appoint a Contracts Manager, who would possibly be responsible for the work should the tender be successful, to act in an advisory capacity during the pre-tender planning and, together with the Estimator and other interested parties, to be responsible for reaching major decisions affecting the use of plant, methods of construction etc.

THE BUYING DEPARTMENT

In order to keep up to date with the current supply situation of materials and the general market trend in prices, the Buying Staff should be constantly available to help and advise in the establishment of realistic prices and competitive quotations.

THE PLANT DEPARTMENT

This department should always be a constant source of benefit when material handling problems involving new equipment are under consideration, and it is quite possible that this Department will be in a position to advise on output standards for the various pieces of plant that are owned by the Company.

THE PLANNING STAFF

In the larger organisation, the Estimator will be able to co-opt the services of the Planning Department in preparing plans for anticipated periods of completion.

During the preparation of the tender, the following matters should be dealt with. These stages in the pre-tender planning technique would be carried out systematically in order to provide the Estimator with information as and when required.

1. The Pre-Tender Report.
2. The Methods Statement.
3. The Plant Schedule.
4. Site Organisation Structure.
5. Schedule of Site Oncosts.
6. Subcontracting arrangements.
7. Pre-tender programme.
8. Estimate Finance Statement.

1. *The Pre-Tender Report*. This should be a comprehensive document containing matters of general interest:

    i General Descriptions etc. (i.e. Directory of people involved.

    ii Site Investigation Report.

    iii Local Conditions in the Area.

    iv Other Factors (i.e. lodging accommodation etc.).

The Pre-tender Report must be concise and to the point. The order of its composition should line up with the contract documents and the contractor's method of estimating. As far as the Builder is concerned, it should be considered an official document and should be properly presented—that is, typed in double spacing, with a wide margin and on one side of good quality paper and contained in a suitable cover bearing the title. It should start with a list of headings dealing with the details of the contract, i.e. the name of the Client, the Architect and the Quantity Surveyor etc. The title page and cover should be set out with a description of the works, the location and correct postal address.

The Report should embrace the results of the detailed site investigation and information regarding local conditions. It is true that too much detail is better than too little; nevertheless, do not provide irrelevant details or give unnecessary advice. Try to avoid any form of ambiguity. Simplicity is essential for clarity. It is sound practice to re-write each paragraph, if necessary, until its meaning cannot be misunderstood.

Site Investigations. It is impossible during estimating to reach decisions concerning the intended methods of executing the work unless the Builder is fully conversant with site conditions. Too much trouble cannot be taken in carrying out a methodical and detailed investigation into the site and local conditions.

It is an excellent arrangement to organise a meeting on site with the Estimator, and a member of the Constructional Staff so that answers to queries can be obtained on the spot. These matters are recorded and form part of the permanent site documents should the tender be successful. In addition, the Estimator should have compiled a list of other questions in the light of his own experience. On the site, his mind should be free to observe, and investigate unusual features. The more complete his check list of points for clarification, the better able he will be to make those special investigations so essential to building up a really useful report.

Many of the questions apply to almost every contract and these might well be printed as a 'Check List for Site Investigation' as illustrated in Section II of the Pre-Tender Report, Figure 3.

Although this document may be considered rather elaborate at the estimating stage, it will prove of invaluable assistance to the Constructional Staff as soon as the job is awarded to the Company.

Figure 3.

PRE-TENDER REPORT

project: Factory and Offices for   reference: P.T.P. 100/59
L. Borrow & Co. Ltd.
prepared by: AJB., DRO and RMcC. date: 13.11.70.

I. Generally
A. client: L. Borrow & Co. Ltd. 607 2871.
B. architect: C. Brown & Partners F/F.R.I.B.A. 708 2991.
C. quantity surveyor: Geo. Clements Esq., A.R.I.C.S. 236 2681.
D. consulting engineer: A. W. Keen Esq., A.M.I. STRUCT. E. 643 9781.
E. description of work: Single storey. 25m × 30m (overall dimensions, etc.)

F. LOCATION (correct postal address): Penforn Road, Grange
    Factory Estate, Bradford 7.
G. ADDRESS OF LOCAL AUTHORITY: Bradford Corporation.
    236 6741.

II. SITE INVESTIGATION

A. ACCESS TO LOCATION BY ROAD, RAIL AND 'BUS SERVICES: Main
    Road A.50 from City. No. 17 Bus.
B. ACCESS TO SITE, CROSSOVERS, TEMPORARY ROADS: Existing Cross-
    over. Suitable but temporary road required 42 m × 2·5 m
    wide.
C. DISTANCE OF SITE FROM MAIN ROADS: On main road.
D. WORKING SPACE AVAILABLE FOR SITING OFFICES, CANTEEN,
    STORES PLANT, MATERIAL, ETC.: Restricted on south side
    only. Adequate on North adjacent to main entrance.
E. SERVICES AVAILABLE TO SITE:
    (i)   Water   ⎫
    (ii) Gas   ⎬ None yet.
    (iii) Electric   ⎭
    (iv) Telephone   Overhead wires across south end of
                         site.
F. CONCEALED SERVICES: Electricity supply to adjacent premises.
G. TRESPASS PRECAUTIONS NECESSARY: Yes. Watchman! Site is
    children's playground.
H. DETAILS OF DAMAGE TO EXISTING STRUCTURES, ETC.: Photographs
    of retaining wall on boundary!
I. NATURE OF GROUND, CONDITIONS OF TRIAL HOLES, ETC. See
    Engineers Report No. 771/28.
J. LOCAL WATER TABLE: Approx. 1·5 m below formation.
K. SITE CLEARANCE DIFFICULTIES (see also METHOD STATEMENT):
    2·1 m cut and fill. Pure clay—hard going.
L. NEAREST TIP: 6 km, £0·20 per m³.
M. INFORMATION ON GRID OF LEVELS: None taken yet by Architect!
N. NEAREST BENCH MARK: Railway bridge abutment 200 m south
    road.
O. PHOTOGRAPHS REQUIRED: (see H. above).

III. LOCAL CONDITIONS

A. AVAILABILITY OF LOCAL LABOUR: Lively building development
    area.
B. ADDRESS OF NEAREST LABOUR EXCHANGE: 40 East View Road,
    Bradford 9. 236 2868.

C. OTHER CONTRACTS IN THE AREA AND APPROXIMATE COMPLETION DATES: School opposite—completion early 1971.

D. INDUSTRIAL COMPETITION: Light engineering only (within 1·5 km radius).

E. OTHER TENDERING: None known at this stage.

IV. OTHER CONDITIONS

A. LODGING ACCOMMODATION: Guest House 1·5 km "The Cedars". Mrs. Jackson. 842 2333.

B. LOCAL HOTELS: 5 km away A.A., R.A.C. 2 star. "Boulton Arms". 242 1242.

C. TRAVELLING TIMES (from City to Site by road, rail, etc.): 15 mins. from Bradford.

D. KEYS TO SITE: Architects office. 0900 hrs.–1730 hrs. Mr. Baker.

E. SPECIALIST SUB-CONTRACTORS IN THE AREA: Plasterer, J. Briggs. High Street. 643 9916. Steelfixer, A. Mathews, Nanchester Road. 236 2167.

F. ADDITIONAL INFORMATION: Plant hire firm—1·5 km away. Yeo, Ltd. 607 2141.

G. RECOMMENDATIONS: Consider appointing J. Smith, Agent who has had similar experience on Walker, Ltd. contract last year.

Signed: AJB.
Date: 13.11.70.

2. *The Methods Statement.*

This is a detailed schedule giving recommendations for methods to be employed for site handling excavation and so on, and is one of the most important matters to be considered at the pre-tender stage. It is usually prepared in connection with the suggested sequence of operations. The type of equipment required for material handling, and many other important matters which will greatly affect the practical approach to the constructional work, must always be determined before any price calculations are carried out. A detailed 'Questionnaire' should be prepared by the Estimator after carefully reading through the Bill of Quantities, outlining all the points for discussion and matters requiring decisions which will affect the price build-up.

The Methods Statement can then be prepared. Each material handling problem should be investigated in detail and several alternatives considered before reaching a final decision. By applying a fundamental principle of method study to this technique, we discover that for most problems several possible solutions can be found, but only ONE solution can be the best. The more thought given to these problems at this stage, the better chance there is of reaching a decision that will enable the most competitive quotation to be submitted. An example of a Method Statement is shown as Figure 4.

Figure 4.

METHOD STATEMENT

I. EXCAVATION

A.	Site Stripping.	Crawler tractor, type 844 (our plant) with 1 m³ (1 m³ = 1.$\frac{1}{3}$ y c approx.) shovel side tipping with 5 m³ tipping wagons.
B.	Reduce Level Digging (cut and fill).	Crawler tractor type E.7 or E.8 (our plant) with 9 m³ scraper unit. 2 machines.
C.	Stanchion Bases.	Hydraulic tractor types JCB (hired). Square hole digging bucket 0·5 m³ capacity.
D.	Foundation Trench and Drains.	Mechanical excavator type 10 RB. (our plant) fitted with back-actor 0·3 m³ capacity. (Load direct into 5 m³ tipping wagons.)
E.	Basement Digging.	Mechanical excavator type 10 RB. with dragline bucket 0·3 m³ capacity. (Allow 60 per cent machine excavation, remainder hand digging.) (Allow banksman.)

II. CONCRETING

A.	Mixer Set-up.	18/12 Reversible drum mixer with built-in weigh batcher and loading shovel. 20 tonne cement silo. Type K.4 Boiler unit de-frosting plant.

III. TRANSPORTING CONCRETE

A.	Foundations and Drains etc. Ground Floor.	1 m³ Dumper trucks.

B. Columns and Beams. Concrete rail transporter with 750 kg hoist tower with concrete skip and automatic roll-over bucket.

C. Floors (suspended). Precast beam unit installed with 10 tonne mobile rubber tyred crane type K.M. 22 (hired).

IV. Reinforcement

Price for "delivered to site cut and bent" in accordance with Bending Schedule (Messrs. Jacksons quotation 18.10.70).

V. Formwork

A. Columns. Standard plywood layout as System C/4/8. (See company's Formwork Manual.)

B. Beams. Standard Wrot Boarded layout as System B/7/12 (ditto).

C. Walls. Metal wall forms layout as System W/8/2 (ditto).

VI. Scaffolding

A. External. Complete independent tubular steel scaffold with staging at 1·8 m intervals around north, south and east elevations only.

B. Internal. Scaffold towers 8·5 m high with access ladder 1·8 m square boarded out platform (4 towers).

VII. Hoisting

A. External Hoists. 750 kg platform hoist for bricks, etc. 750 kg hoist with skip attachment. (See Section III, B.)

B. Lightweight Scaffold Hoists. Small petrol driven "Jace" scaffold hoist for mortar and small hoisting requirements.

VIII. Miscellaneous

A. Pumping. Two 75 mm diameter submersible pumps to be included during excavation. (Water Table only 1·5 m)

40

B.   Mortar Mill.   One pan-type mortar mill for brick-
                             laying together with concrete hard-
                             standing.

                             Signed: E.O.H.
                             Date: 13.11.70.

3. *Plant Schedule*

After the preparation of the Methods Statement, a summary
of the plant requirements for the project should be detailed,
giving a schedule of mechanical equipment intended for use
on the project. This summary must show the type of equipment required, any ancillary equipment and the anticipated
period of hire. It should also include details of suggested plant
maintenance arrangements on Site for the larger building and
civil engineering projects.

It is of great advantage for the Estimator to have a complete
list available of the company's mechanical plant. This will
prove of great assistance as a *Check List* to make sure that
nothing has been overlooked during the preparation of the
summary of Plant requirements. An example of a Plant
Schedule is shown as Figure 5.

4. *Site Organisation Structure*

In order that sufficient allowance is made in the estimate to
cover the administrative costs on site, a proposed organisation
structure should be prepared to show the lines of authority
and the delegation of responsibilities from the Site Manager, or
Agent, down to the supervising grades of Trades Foremen and
Chargehands. The amount of time allocated for supervision
by Trades Foremen and Chargehands should be expressed as a
percentage of their weekly wage. A typical example of a Site
Organisation Chart is shown as Figure 6. This information, read
in conjunction with the Pre-Tender Programme, will give a true
and realistic guide when calculating the estimated site overheads.

At certain stages of a Contract, the greater proportion of
work may be carried out by Sub-Contract Trades, but this
may not reduce the amount of Supervision by the main
Contractor needed on the Site. It is therefore most unwise to
express the value of Site Supervision as a percentage of the value
of Builders Work. Only by a careful and analytical approach

to Site Administration at this Pre-tender stage will a realistic allowance be made in the estimate.

5. *Schedule of Site Oncosts*

The purpose of Site Oncost,[1] or Site Overheads Analysis, at the tender stage is to provide the Estimator with a realistic, practical assessment of preliminary items and overhead allowances required for the efficient and effective execution of the work. It has already been explained how the build-up of the site administration is established. This assessment of site personnel, in conjunction with other matters such as site accommodation, extent of site offices, stores, Watchman's facilities, hoardings etc., contributes towards the allocation of realistic Site Overheads.

A Check List of items for consideration must be used, similar to the information contained on the Pro-Forma, Figure 7. This document, when properly completed by the Estimator, will show the precise build-up of the Site Overheads allowed in the estimate.

6. *Sub-Contracting Arrangements*

Schedules should be prepared of all trades intended for subletting and arrangements made to send out full and comprehensive enquiries to all Sub-Contractors from whom the Builder

Figure 5.

PLANT SCHEDULE

No.	Description	Weeks on Site	Remarks
1	Crawler tractor Type 844 1 m³ shovel (side tipping).	2	Our plant.
2	Crawler tractor Type E.8 with scraper unit 9 m³ capacity.	3	1 our plant, 1 hire.
1	Hydraulic tractor Type JCB 0.5 m³ bucket attachment.	3	Hired.
1	Mechanical excavator Type 10 RB. bucket attachment.	4	Our plant.

[1] John C. Maxwell-Cooke, *Civil Engineering Contracts Organisation*. Cleaver- Hume Press Ltd. p. 55.

42

1	Mechanical excavator Type 10 RB. 0·3 m³ backactor and 0·3 m³ dragline attachment.	4	Our plant.
1	18/12 reversible drum mixer (electric) Type B/11.	30	Consider purchase or hire.
1	20 tonne cement silo	30	Our plant.
1	Type K.4 boiler de-frosting unit	30	Our plant.
2	1 m³ rubber tyred dumper truck.	30	Our plant.
1	set Concrete rail transporter unit with 200 m track.	30	Our plant.
1	set 30° points plus 12 slow bends.		
1	10 tonne mobile crane Type KM.22	1	Hired (Messrs. Corton, 236 9211)
2	750 kg hoists. 1 platform attachment. 1 concrete skip attachment.	15	Our plant.
1	Lightweight scaffold hoist.	12	Hire.
2	75 mm diameter pumps. Type S/11/8. 30 m hose to each.	8	Hire.
1	Mortar pan-type mill.	20	Our plant.

requires quotations. A full investigation into market trends at this stage and complete coverage of all suitable Sub-Contractors will not only reduce excessive clerical work in sending out further enquiries should the tender be successful, but it will also enable the firm to base their overall quotation on more competitive Sub-Contractors' prices. This procedure may not prevent further negotiations with Sub-Contractors should the Builder be successful in obtaining the Contract. Of course, in certain circumstances, it would be unethical to engage in further negotiations with Sub-Contractors after the Contract has been obtained, unless it had been made clear at the time of their original tender that further negotiations would take place.

7. *Pre-Tender Programme*

It is becoming increasingly evident that contracts are awarded to Builders who pledge to complete the work in a short time. Other things being equal, should the price be reasonable, the job is often awarded on a basis of time rather than money. Considerable thought must, therefore, be given to the

Figure 6.

A TYPICAL PRE-TENDER SITE ORGANISATION STRUCTURE

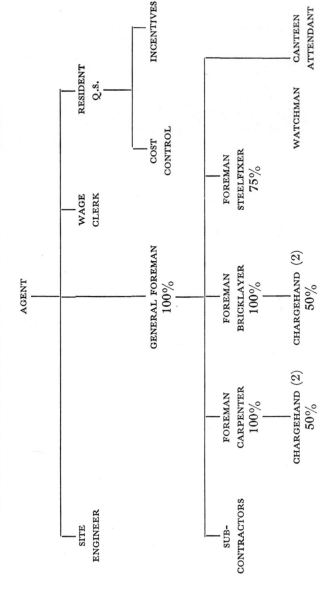

Percentage indicates amount of supervision.

period of completion of the contract. Every attempt should be made to determine the client's requirements and the contractor may have to ascertain whether *time* or *price* will be the main consideration before being able to submit a competitive tender. Each operation involved in the contract must be considered on its merit—the broad outline showing the phasing of operations of the Specialist's work carefully investigated. An example of a Pre-Tender Programme is shown as Figure 8.

Figure 7.

SCHEDULE OF SITE ON-COSTS

PROJECT: Factory and Offices for   REFERENCE: P.T.P. 100/59.
        L. Borrow & Co. Ltd.   DATE: 13.11.70.
PREPARED BY: AJB., DRO. and RMcC.
CONTRACT PERIOD: 50 weeks.

I. SITE PERSONNEL

No.	Description	Weeks on Site	Remarks
1	Agent	50	
1	Deputy Agent	50	
1	General Foreman	50	Mr. Smith if possible.
1	Deputy General Foreman	50	
1	Engineer	26	
1	Assistant Engineer	26	Early stages only.
1	Junior Engineer	10	
1	Resident Senior Surveyor	50	
1	Assistant Surveyor	26	Early stages only.
1	Junior Surveyor	10	
1	Time Keeper	50	
1	Clerk ⎫ Cashier ⎭	50	
—	Assistant Cashier		
	Storekeeper ⎫		
1	Canteen Attendant ⎬	50	Mr. Brown lives near.
	Watchman ⎭		
—	Traffic Controller	—	
1	First Aid Attendant	50	
1	Fitter	26	

45

II. Site Accommodation

No.	Description	Weeks on Site	Remarks
1	Agent's Office	50	Fully equipped.
1	Foreman's Double Office	50	Fully equipped.
1	Foreman's Single Office	50	Fully equipped.
1	Engineer's Office	26	Fully equipped.
1	General Office.	50	Fully equipped.
1	Time Keeper's Office	50	Fully equipped.
1	Clerk of Works' Office	50	Including GPO phone.
1	Canteen	50	
1	Stores	50	
1	Watchman's Hut	50	
3	Latrine Huts	50	
1	First Aid Hut	50	
1	Canvas Shelter	40	
2	Cement Sheds	28	
1	Storage Compounds (25 m × 30 m)	40	
1	Hoarding (150 m)	50	
1	Fencing (75 m)	50	
1	Fitter's Shop	26	Early stages only.
1	Carpenter's Shop	30	
1	Inflammable Store	30	
1	Painter's Hut ⎱ Plumber's Hut ⎰	26	

III. Generally

	Description	Remarks
i	Period required for pre-contract planning.	6 weeks.
ii	Subsistence allowance.	£3·25 (if applicable).
iii	Out of town allowance for Agent/General Foreman	£5·00 plus car. (Agreed by M.D.)
iv	Unloading of materials.	None.
v	Hard standings.	For mixer set-up 18/12.
vi	Inclement weather precautions	None.
vii	Overtime allowances.	5 hours per man per week for 20 weeks.
viii	Floodlighting.	None.

46

IV. SPECIAL RECOMMENDATIONS FOR SITE OVERHEAD ALLOWANCES

An experimental system of Site Costing will be employed during the construction of the *frame only*. One week's training course will be arranged for the senior site staff. Cost of training course fee £10·50 per head (discuss this with M.D. to determine if "site on-cost").

> *Signed:* AJB.
> *Date:* 13.11.70.

Care must be taken not to assume that periods of completion are likely to be in proportion to the total value of the contract. This practice is misleading and will, in the light of present-day conditions on building sites, often be grossly unrealistic. It may well be that a contract of a value exceeding £100 000 could be completed far more quickly than a contract of much less value, it is possible that the operations involved in the smaller contract were of a more complex nature than those in the larger.

It is extremely difficult to anticipate the extent of time to be taken up by Specialist trades at the pre-tender planning stage. Every effort should be made to obtain from the Architect the names of the important nominated Specialists, for example, structural steelwork. A preliminary discussion may then take place to establish a reasonable and realistic time for the completion of the key Specialists trades which would largely determine the duration of the contract.

The broad view of the project is provided by the Pre-tender Programme which must embrace all major operations, showing the period of completion for each separately, as well as their realistic relationship one with the other. This will provide a Sequence Chart based on the information available from the Methods Statement, and the Pre-tender Report. It is then possible to calculate the Plant Schedule and the Schedule of Site Oncosts from this programme.

8. *Estimate Finance Statement*

This statement is a simplified summary and financial breakdown of the total costs anticipated for the project and involves the following:

Figure 8.

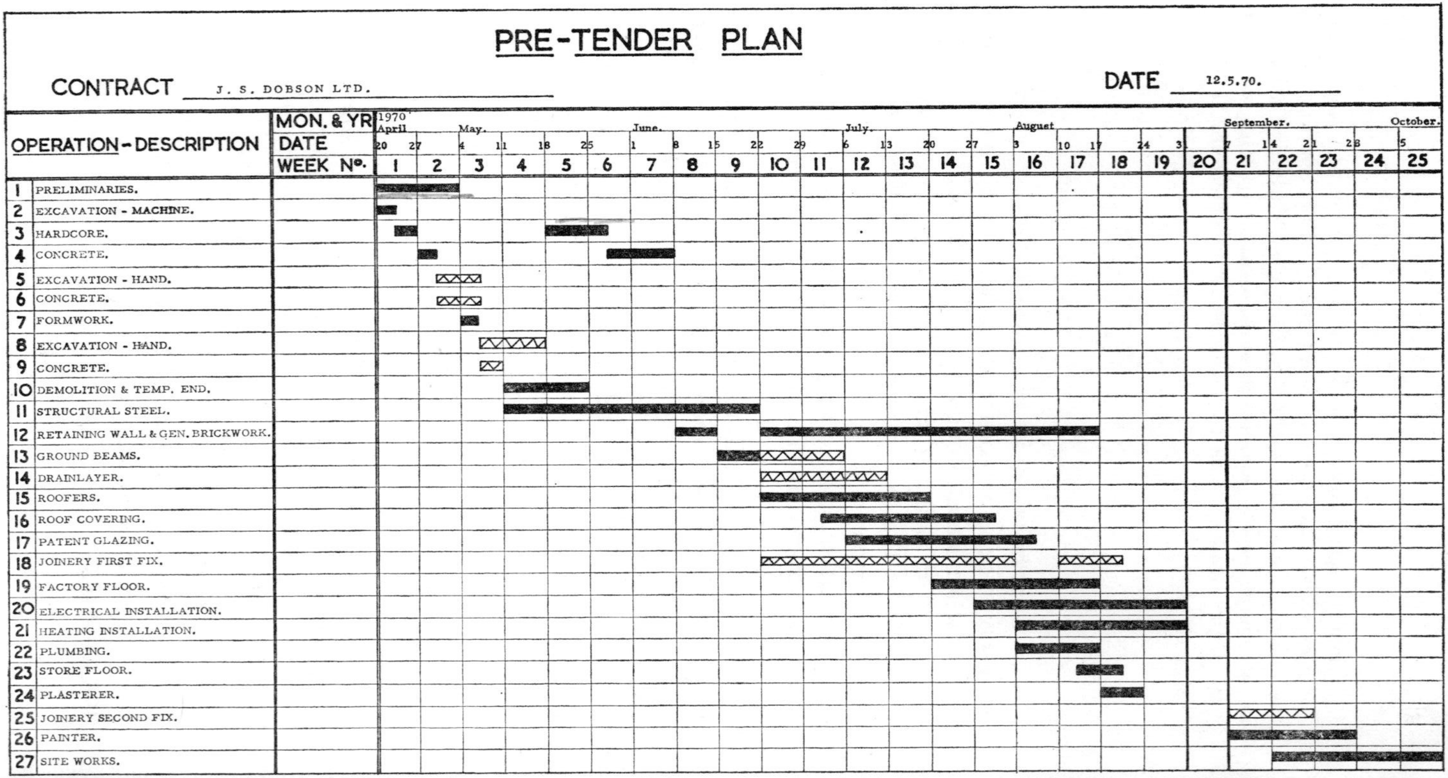
PRE-TENDER PLAN
CONTRACT J. S. DOBSON LTD.
DATE 12.5.70.
OPERATION - DESCRIPTION
MON. & YR
DATE
WEEK No.
1970 April
May.
June.
July.
August
September.
October.
20 27 4 11 18 25 1 8 15 22 29 6 13 20 27 3 10 17 24 31 7 14 21 28 5
1 2 3 4 5 6 7 8 9 10 11 12 13 14 15 16 17 18 19 20 21 22 23 24 25
1 PRELIMINARIES.
2 EXCAVATION - MACHINE.
3 HARDCORE.
4 CONCRETE.
5 EXCAVATION - HAND.
6 CONCRETE.
7 FORMWORK.
8 EXCAVATION - HAND.
9 CONCRETE.
10 DEMOLITION & TEMP. END.
11 STRUCTURAL STEEL.
12 RETAINING WALL & GEN. BRICKWORK.
13 GROUND BEAMS.
14 DRAINLAYER.
15 ROOFERS.
16 ROOF COVERING.
17 PATENT GLAZING.
18 JOINERY FIRST FIX.
19 FACTORY FLOOR.
20 ELECTRICAL INSTALLATION.
21 HEATING INSTALLATION.
22 PLUMBING.
23 STORE FLOOR.
24 PLASTERER.
25 JOINERY SECOND FIX.
26 PAINTER.
27 SITE WORKS.

(a) Preliminaries and Site Overheads.
(b) Labour Cost.
(c) Material Cost.
(d) Mechanical Plant and Transport Costs.
(e) Sums to be included for Nominated Sub-Contractors and Suppliers.
(f) Sums to be included for the Builders' own Sub-Contractors.
(g) Provisions and Contingencies.

The summary sheet should include calculations for the establishment of overhead allowances based strictly on the Pre-Tender Report and Schedule of Site Overheads. This Statement would then be presented to the Management for consideration and final approval before the tender figure is submitted. An example of an Estimate Finance Statement is shown as Figure 9.

Accurate planning at the pre-tender stage, based on systematic research, will reduce the time and effort needed after the contract has been obtained and increases the effectiveness of supervision and control of the work when in progress. No benefit can be obtained from divorcing the estimating side of a company from the constructional staff at the tender stage. Only by a concerted effort to pool the firm's resources and experience can maximum benefit be obtained and accurate, realistic and competitive quotations be prepared.

Figure 9.

ESTIMATE FINANCE STATEMENT

PROJECT: Factory and Offices for
      L. Borrow & Co., Ltd.
NET LABOUR: £45 000.

DATE: 13.11.70.

	£	£
Bill of Quantities		700 520
Less: Provisions	600	
Contingencies	2 500	
P.C.'s	125 605	
Sub-Contractors	300 550	429 260
		271 260

D

Builders own work.

On-Costs:			
	Staff	13 000	
	Phone	208	
	Offices	750	
	Water for Works	800	
	Insurances	11 500	
	Scaffold	4 000	
	Special Plant (hoists, cranes, etc.)	3 000	
	Travel Expenses	1 700	
	Fixed Price Allow.		
	Labour	4 000	
	Materials	5 000	43 958

	315 218
Overheads and Profit 10%	31 521
P.C.'s	125 605
Profit 2½%	3 140
Sub-Contractors	300 555
Profit 2½%	7 514
Provisions	600
Contingencies	2 500

TOTAL OF TENDER £786 653

CHAPTER IV

PRE-CONTRACT PLANNING—PART I

'The determination of the path which will result in the greatest economy of motion and the greatest increase of output is a subject for the closest investigation and the most scientific determination.'   *Frank B. Gilbreth.*

In the previous chapter we considered the importance of systematic research and investigation into the information available at the tendering stage, to enable the estimator to submit a realistic and competitive quotation. If full advantage is to be gained from this, then the successful contractor must further develop and expand the investigation to cover the work stage by stage in detail, before actual building operations begin.

Failure to plan properly before work starts invariably delays completion and involves heavy additional cost. Most of these costs were unforeseen at the estimating stage and in consequence are usually passed on to the client for payment.

Upon being awarded the contract, the builder should be provided with the opportunity to plan the work in advance.

In many cases, builders will encounter difficulty in obtaining complete co-operation and understanding from the architect and client, who are often impatient and unaware of the expense that a hurried start inevitably causes.

The wise client will appreciate the importance of allowing time for systematic fore-thought. The competent Architect should guide the client and expect the builder, after studying a set of full working drawings, to place sub-contracts, order materials, arrange for plant and equipment, engage labour and prepare his schedule of operations and the general organisation and programming of the work.

This is the ideal situation, which builders rarely enjoy. In consequence to what extent the builder will be prepared to sacrifice construction time to spend a few days or weeks of the contract period in thinking ahead before launching a full-scale offensive on the site will depend upon the size and nature of work.

GENERAL PRINCIPLES OF PRE-CONTRACT PLANNING

Any system of overall pre-contract planning should provide an opportunity for supervisory site staff to consider carefully the amount of work to be undertaken, the methods to be adopted and the timing of its execution.

As a general guide to the preparation of a programme, the contractor and his supervisory staff need the following information:

1. A broad picture of the work showing the main operations and their sequence.
2. How long each operation will last.
3. The estimated labour strength required in order to complete each operation as planned.
4. The type of plant and output required to achieve the time period for each operation.
5. The type and amount of material needed for each operation, and the latest date for delivery.
6. The details of specialists' work insofar as it will affect the progress of the builder's own operations. The time period required to enable the specialists to complete their work.
7. The latest date when outstanding information, detailed drawings and specifications or samples approved must be finalised in order to ensure continuous and smooth running of all operations (specialist work as well as main contractor).
8. Provision to record actual progress in relation to anticipated or planned progress.

It will be appreciated that a large proportion of this information is difficult to obtain at the outset of any contract. It does, therefore, call for close teamwork on the part of those principally concerned—the client and architect, to settle design details; the Quantity Surveyor, to provide accurate quantities and reliable information about the Nominated Sub-Contractors and Suppliers. The Contractor's problem of obtaining co-operation from them all is a challenge to the finest qualities of organising talent.

RELATION TO PRE-TENDER PLANNING

All matters that are considered at the tender stage and which affect the price build-up of labour and material and plant must be analysed further before commencement of work on site.

We have already dealt with the various stages of pre-tender planning; where these techniques have been undertaken the estimating staff will furnish the following information:

1. Pre-tender Report.
2. Methods Statement.
3. The Plant Schedule.
4. Site Organisation Structure.
5. Schedule of Site Oncosts.
6. Sub-contracting arrangements.
7. Pre-tender Programme.
8. Estimate Finance Statement.

The purpose of overall planning is to analyse and further consider all this, in the light of any additional information that may be available.

PROCEDURE UPON AWARD OF THE CONTRACT

Upon being awarded the contract, the top management will consider the appointment of suitably experienced and trained persons to take up the responsibilities of Resident Site Agent, General Foreman and so on.

The actual titles of the positions will be purely a question of company policy. On extremely large contracts, a Resident Projects Manager may be appointed.

In addition, the Contracts Manager or Supervisor (usually based at Head Office) will be appointed, together with the allocation of responsibilities to other Head Office personnel, such as Surveyor, etc.

In larger organisations, the Planning Department, which receives all documents and details directly from the Estimating Department may, under instructions from Board level, convene a meeting of all concerned.

This preliminary meeting is really intended to announce the

award of the contract and give everyone a broad appreciation of what is involved. When a formal summons concerning the meeting is circulated, it is often attached to a detailed information sheet outlining brief particulars regarding address of job; value of contract; staff appointments to the contract, etc.

A copy of a typical New Contract Information Sheet is attached as Figure 10.

THE PRELIMINARY PLANNING MEETING

After the initial announcement of the award it is important that a systematic approach to the preliminary discussion is adopted.

Figure 10.

NEW CONTRACT—INFORMATION SHEET

DATE: 3rd May, 1971.   JOB NO.: 2171.

1. CLIENT: L. Borrow & Co., Ltd.
2. CLIENT'S ADDRESS: Penform Road, Grange Factory Estate, Bradford.
3. ADDRESS OF JOB: As above.
4. VALUE OF CONTRACT: £1 000 000.
5. CONTRACT PERIOD: 12 months.
6. ANTICIPATED COMMENCEMENT DATE: 12th June, 1971.
7. DESCRIPTION OF WORK: Single storey factory. Overall dimensions 245 m × 30 m with additional ancillary work including roads, etc.
8. ARCHITECT: C. Brown & Partners F./F.R.I.B.A. 607 2991.
9. QUANTITY SURVEYOR: Geo. Clements, A.R.I.C.S. 236 2681.
10. CONSULTING ENGINEER: A. W. Keen, A.M.I.STRUCT.E.
11. PRODUCTION MANAGER: I. Kemble (Ian).
12. SURVEYOR: W. Charlton (Bill).
13. AGENT: J. Frost (Jack).
14. GENERAL FOREMAN: Not yet appointed.

One of the first items on the agenda must be an investigation into the information available. Upon the extent of information will often depend the *practical* commencement of the work. In consequence a simple system of checking over all relevant details, specifications, letters of instructions etc. will establish what further information is needed for detailed planning purposes.

At this stage, a little foresight will enable the contractor to agree to contractural starting dates, dependent upon sufficient information being available. A simple check list can be prepared, based on experience, to determine the absolute practical *minimum* information needed in order to consider setting the wheels of the builder's organisation in motion.

A typical list of minimum requirements might be similar to the following:

1. Fully priced Bill of Quantities.
2. Schedule of Sub-Contractors.
3. Copies of all Enquiries and Estimates.
4. 1:100 Scale Plan—All Floors.
5. 1:100 Scale Elevations and Sections Block Plan.
6. Specialist details (i.e. Bending Schedules).

If a decision is reached at this stage that insufficient information is available, then arrangements to start detailed planning or commencement of work on site can be delayed. A prompt communication to the architect outlining the information that is required will substantiate at this stage a claim for adequate details which are required in order to undertake the work. This type of letter should become a routine in the procedure adopted at the outset of planning, and a 'standard' letter can be prepared for this purpose. It can then be used periodically during the later stages of planning whenever details or information is needed from the architect or engineer.

REGISTRATION OF DRAWINGS

Architect's details and other specialists' drawings arrive from time to time and should be recorded systematically in a Drawing Register. Provision in the Register should be made to record.

1. Title of Drawing or Description.
2. Drawing Ref. Number.
3. From whom received.
4. Main Scale.
5. Date Received.
6. No. of copies received.
7. Date distributed within the company.
8. Details of distribution.
9. Provision for details of subsequent amendments.

Figure 11.

DRAWING REGISTER

CONTRACT _____

CONTRACT Nº. _____

DRAWING Nº.	TITLE OR DESCRIPTION	MAIN SCALE	RECEIVED FROM	ORIGINAL		AMENDMENTS				DISTRIBUTION
				Nº.	DATE	A	B	C	D	

Each drawing is then usually stamped to identify that it has been properly registered. A copy of a typical page of a drawing register is shown as Figure 11. Two copies of this register can then be kept up to date; one copy at head office and the other retained on Site.

It will often be experienced that 'preliminary' drawings will be sent to the contractor at an early stage. As a precaution against errors occurring from the use of these preliminary details, it is wise to print 'PRELIMINARY—NOT A WORK-ING DRAWING' in large letters across the face of these drawings. A large purpose-made rubber stamp is a useful piece of equipment to have for this purpose. At a later date, when revised or amended details are issued an official rubber stamp entitled 'THIS DRAWING HAS BEEN SUPERSEDED' should be imprinted in a conspicuous place on the drawing to avoid errors caused by working to obsolete details.

THE BILL OF QUANTITIES

The Bill of Quantities plays a very important part in planning the work. For planning purposes the Bill represents an analysis of each and every operation involved in the complete construction. When priced, it constitutes the buildup of the tender but to the planner and the construction staff it is more than a quotation—it becomes a schedule setting out the estimated cost for labour, plant and material, amongst other things, for each unit of work to be carried out. In order to provide an accurate and reliable basis upon which to base a programme of work, the priced bill or bill rate must be broken down into more detail. We are concerned with three main factors in calculating a time period for an operation:

1. Labour output or cost.
2. Plant output or cost.
3. Site supervision and site overhead requirement or cost.

In consequence, the traditional method of pricing a bill of quantities showing a 'bulk rate' for each item is insufficient to provide the planner with a basis for his calculations.

It is becoming increasingly common for the estimator to show the breakdown of his unit rates by sub-division into separate columns (shown in pencil on the back of the preceding page or

Figure 12.

A TYPICAL BREAKDOWN OF THE ESTIMATE

Ref.	Description	Qty.	Unit	Labour £	Plant p	Material £	O/H × Profit £	Total Rate £	£
A	65 mm Common Bricks P.C. £12·50 per thousand delivered to site, in cement mortar as described:								
B	Reduced brickwork (to one brick thick)	41	m²	1·74	2½	1·76½	0·53	4·06	166·46
C	Half brick wall	105	m²	0·87	1½	0·88½	0·27	2·04	214·20
D	275 mm Hollow wall of two half brick skins, with wall ties as described	206	m²	1·79	2½	1·78½	0·54	4·14	852·84
E	Extra on brickwork for fair face and flush pointing	53	m²	0·15	—	—	0·02	0·17	9·01
F	Damp proof course to chimney stacks above roof of two courses Welsh slates laid to break joint and bedded and floated in cement mortar (1:2)	5	m²	1·63	—	0·81	0·36	2·80	14·00
G	Render 12·5 mm thick in cement mortar (1:3) on chimney stack in floor and roof space as the work proceeds	6	m²	0·24	1	0·09	0·05	0·39	2·34
H	Bitumen damp course to B.S. 743 Type 5A., 381 mm girth, as cavity gutter, and seal edge with mastic	60	m	0·13½	—	0·20	0·04½	0·38	22·80
J	Bed corrugated asbestos cement sheeting to half brick at eaves or verge in cement mortar and point both sides	48	m	0·16	—	0·04	0·03	0·23	11·04
K	Close 50 mm cavity in hollow wall under sill with slate pieces in cement mortar	89	m	0·16	—	0·09½	0·04	0·29½	26·25½
L	Close 50 mm cavity in hollow wall at reveal with brickwork bonded to outer skin and two courses Welsh slates 115 mm wide in cement mortar (1:2)	189	m	0·16	—	0·17	0·05	0·38	71·82
M	Labour setting forward outer skin of hollow wall for a width of 228 mm, 57 mm from general face at quoin, including cuttings	33	m	0·16	—	0·05½	0·03	0·24½	8·08½
									1398·85

by superimposing a stick-on slip on the rate column in the bill). An illustration of this is shown as Figure 12. A simpler method is to price labour only rates in pencil on the extreme left hand side of each page against the reference or quantity column. Whichever method is used, it is important that the results of the estimator's calculations for actual cost of labour and plant are clearly shown. An item could be included on the agenda to check this at the preliminary planning meeting. If the bill has not been broken down into those basic rates, arrangements must be made for it to be done at an early date.

Whilst considering the Bill, additional copies should be requested as required. It is useful to have a spare bill available during planning in order to make notes against items to guide and assist the site at a later date. Providing the site with a fully priced bill of quantities is a controversial point and is a side issue that will not be discussed here. Nevertheless, some form of schedule, priced bill or typed sheets with bill reference and labour and plant rates must be made available for use on the site as and when required.

RELEVANT CORRESPONDENCE

Many letters pass between Architect and Builder and Quantity Surveyor and Builder from the date of acceptance and before work actually commences on site. Often instructions about Nominated Sub-contractors and Suppliers and amendments to Bill of Quantities and Specification are issued during this period. At an early stage in the planning procedure, all correspondence should be read through and duplicate copies made of any relevant letters to or from Head Office for retention on the site. It is unlikely at this stage that typing or clerical staff will be familiar with the new job and in order to prevent any letters going astray a folder should be issued and retained by the planning team to contain all site copies of letters etc. until the work commences.

ADDITIONAL ITEMS OF ADMINISTRATION

At this early stage in the arrangements for getting the contract underway, a great many small details indirectly associated with the construction work have to be arranged. Contact with

Local Authorities is most important and should be made well in advance of commencing work on site.

By research into the company's experience of starting new jobs, it is possible to prepare a very detailed list of almost all the items requiring immediate attention—laying on water supply, gas, electricity, etc. To spend time going through these lists systematically will ensure that important arrangements are dealt with promptly and thereby provide Local Authorities' services, etc. to the site right at the beginning of the job. Taking these items together early on in the planning stage will leave the site representative free to devote his thoughts to detailed construction problems. We will now consider some of these points in greater detail.

INSURANCES

The contractor's obligations as far as insurance is concerned will be covered by the R.I.B.A. form of contract, Clause 15. Care must be taken to check over those legal obligations in order to make sure that adequate coverage has been obtained with the appropriate insurance company. In addition to the formal policies covering fire risk, etc. it is advisable for the builder to look carefully into the question of *additional* insurance. Often difficult excavations may involve substantial risk and it may be considered worthwhile to invest a few pounds in obtaining a cover against damage caused by collapse of earthworks. If the contractor has any demolition work to undertake, care must be taken to check that the insurance policy covers taking down as well as erection of buildings. Modern designs for front elevations frequently involve large areas of glass; the contractor is often faced with considerable replacement costs due to vandalism or accident during the construction. Insurance cover in this respect may often prove to be an investment.

There are very few things likely to go wrong on a building site that an insurance company will not cover if the premium is large enough. It is up to the contractor to exercise foresight in this direction and reduce possible losses by insurance before work commences.

WATER FOR THE WORKS

Supply of water to the site on most building contracts is

essential and any delay in obtaining adequate supply will cause a disruption of work right at the commencement of the job. Applications to the local Water Undertaking should be made very soon after the award of the contract. It is a great benefit if a small sketch plan is submitted with the application showing the site and the required position of entry of the water supply in relation to site accommodation, mixer set-ups, stand pipes and the like. Before work commences a water fee has to be paid. The amount varies according to locality but is usually 'so many p' per £100·00 of the contract value.

ELECTRICITY FOR THE WORKS

With the rapid increase in the use of more modern plant, it is necessary to obtain an electric supply to provide power for electrically driven plant such as mixers, hoists, cranes etc. The plant requirements must be carefully thought out at this stage and the load calculated in order that adequate services are installed. The installation of a temporary electric supply is often extremely costly and in consequence every effort should be made to arrange for the main permanent supply for the completed building to be installed. The cable can be brought within the boundary of the site and the contractor can 'tap off' by temporary meter for the builder's electricity supply. Prompt action in arranging this facility will often save the builder considerable cost in temporary electrical installations.

GAS SUPPLY

When dealing with the main supply for the completed building any gas service required by the client should be considered. Services may then be brought in at the foundation stages in the erection of the building and will save cutting and chasing for pipe runs during finishing.

TELEPHONES

Prompt communcation between the site and Head Office is essential, particularly at the start of a job. Often, after the written application for a site telephone has been made, the contractor has to wait several weeks while the General Post Office make the necessary surveys to establish whether or not a line is available. This delay usually causes embarrassment to all concerned. It is important therefore that the time the site is without a telephone should be reduced to a minimum. The application, together with a sketch of the site showing location of office accommodation should be despatched to the G.P.O. without delay. Where large contracts are undertaken, contractors may install alternative means of communication with Head Office such as short wave radio over short distances. Teleprinter or the G.P.O. 'Telex' service is also available where the cost and distance justifies the installation.

On certain types of work, large housing estates, multi-storey work etc. it is possible that there will be several site offices located in different parts of the site. In these circumstances, an internal telephone, usually battery operated ex-W.D. equipment, is an ideal method of getting in touch quickly with site staff.

LICENCES AND PERMITS

Where the contractor requires access to the site by crossing a public right of way such as a pavement, a licence is required by the Local Authority. This also applies when hoardings or gantries are required adjacent to the public highway. The contractor must make formal application and pay the licensing fees for these documents at an early date and must make sure that a copy is retained at the site before work commences.

CONNECTIONS TO THE MAIN SEWER

Drainage connections to main outfalls is often an important *immediate* consideration. Where connections are required to Main Sewers, an application has to be made to the Local Authority for permission to discharge effluent. In some localities, the Public Works Department actually undertakes the sewer connection themselves. In this case the builder must obtain a quotation for the work if one has not already been submitted to the Architect. Usually the Authorities require several weeks' notice. The builder must establish the date when the sewer connection is required and notify the Authority in writing well in advance. This should be accompanied with sketch of site showing proposed drain runs and sewer connections.

COMMENCEMENT NOTICES

The Local Authority for the area where the work is to be undertaken will require formal notification that work is about to commence. This must be done at an early date before actually starting on site. The Factory Acts requires the builder to submit Form 10 not later than seven days after the beginning of operations on site to the Factory Inspector for the district. The form contains brief particulars about the nature of the work, whether mechanical power is to be used and so on. It is advisable to deal with this as soon as possible after the award of the contract.

THE PROBLEM OF CONCEALED SERVICES

Frequently the contractor will encounter existing underground services running across the site—water, gas and electricity—which are providing a service to adjacent premises and these are often not shown on detailed drawings of the site. The builder is well advised to write to all authorities:

Gas Department
Water Department
Electricity Authority
General Post Office

and obtain written confirmation that the site upon which the contractor is to commence work is clear of concealed services which may become highly dangerous if damaged and disrupt the excavation programme.

At a preliminary planning meeting all these items should be carefully considered. A member of the team should be delegated to deal with these matters and a check made just before starting work on site to make sure that everything has been properly 'laid on'. A systematic approach to these details will ensure that nothing that will hold the job up in its early stages is forgotten.

CHAPTER V

PRE-CONTRACT PLANNING—PART II

'Thinkers of today recognise that the work to be done is so great that . . . . co-operation and teamwork is the crying need.' *W. R. Spriegal and Clark E. Myers.*

The Contractor must now turn his attention to the site conditions. If he has carried out a proper survey at the estimating stage he will be able to refer to the Site Investigation Report—described in Chapter III, page 35.

We will now consider in greater detail the various points associated with site conditions which should be carefully noted before commencement of operations.

LOCALITY

In order that correspondence is properly directed to the site office the correct postal address must be established. If there is some difficulty in finding the right address, a visit to the local post office will settle the problem.

The nearest available public transport with the particulars of the terminus in the nearest city will help to direct new labour to the site.

ACCESS

Perhaps the most important fact—how do we get into the site? The contractor must measure widths and heights of openings on the obvious routes to ensure that lorries and equipment will be able to negotiate them. If there is no made access road to the site, a note of the dimensions and other requirements should be made of any temporary roadwork necessary. In some cases special quotations will have to be obtained from Subcontractors or Suppliers if the site is above a certain distance from a main 'hard road'. This must be borne in mind and the critical dimensions checked.

POSITION OF EXISTING SERVICES

Where overhead cables and other services cross the site they must be plotted on a plan and any overhead service that requires re-positioning should be notified to the appropriate authority at an early date. The problem of concealed services has been dealt with in the preceding chapter.

The position of the nearest public telephone is worth noting and, of course, the telephone number. Where it is anticipated that difficulty will be encountered in obtaining a direct telephone line from the G.P.O., arrangements might be made, with the co-operation of local tenants to record their telephone number for use in an emergency.

Existing water supplies must be investigated and any hydrant points duly noted on a plan of the site.

NATURE OF THE GROUND

This item appears in the preambles of most Bills of Quantities and is often disregarded at the estimating stage. It is most important that a proper study of sub-soil conditions should be made *before* excavating method or plant is decided upon. In many cases, where the Architect anticipates that the ground condition will affect the design of foundations, trial holes are made and suitable reports are available for scrutiny by the contractor. Where no trial holes exist it is a wise precaution for the contractor to excavate them at his own expense at selected points around the site to satisfy himself about the condition of the ground at lower levels.

STORAGE FACILITIES

Availability of sufficient storage space, and areas for the installation of offices, etc. must be carefully taken into consideration. On restricted sites, the possibility of storage on adjacent plots of ground should be investigated. Where space is limited, the installation of multi-storey office accommodation or offices elevated on scaffold gantries may be the only solution.

TRESPASS PRECAUTIONS

Some areas, particularly city sites, are vulnerable to trespass and vandalism. This must be an important consideration. Hoarding and fencing is often measured in the Bill of Quantities.

E

Whether or not the Quantity Surveyor has included this item, the contractor must, in his own interests, protect his site and material. It is now possible in some areas to call upon companies who will provide the services of guard dogs with skilled handler, to act as a worthwhile deterent against unlawful entry on the site out of working hours. The builder may consider the appointment of a night watchman who may also be well worth while particularly in areas over-run with children.

ADJACENT BUILDINGS

Before excavation work begins, a careful check must be made of the condition of buildings and walls adjacent to the site. Settlement often takes place when excavation work is carried out up to the face of existing walls. More often than not, adjoining tenants become aware (often for the first time) that cracks exist in their own building. Photographs taken before work started, properly dated and showing the cracks visible on walls overlooking the site will protect the builder against unfounded claims for damage to existing adjoining property. Glass 'tell-tale' slips can be placed across existing cracks to ensure that any additional settlement is noticed and properly recorded.

BENCH MARKS

Local Ordnance Survey bench marks which can be used in preparing a grid of levels, to check reduced level excavations and floor levels, should be plotted on a plan of the site and the levels recorded.

TIPPING FACILITIES

If the excavation includes the removal from site of large quantities of soil, knowing the position of the nearest tip will assist the builder to calculate the number of lorries required to maintain maximum output from mechanical excavating equipment. If a tipping fee is involved, this should also be determined at this stage.

PHOTOGRAPHS

Exceptional site conditions, such as bad ground, liability to flooding etc. should be recorded by photographs before the builder commences work. Site conditions may change due to bad weather and the Quantity Surveyor's impression of excavation work at the stage of preparing the Bill of Quantities may not coincide with actual ground conditions at the start of the job. Photographs are an important part of the Builder's records in these circumstances and should be carefully preserved for future reference.

SITE LAYOUT DRAWINGS

The Architect will usually provide a block plan showing the proposed new building in relation to adjoining premises, roads etc. This plan is usually drawn to a scale of 1/500 or 1/2500. This is far too small to be of any use to the builder when considering the layout of his office accommodation, mixer set-up, storage positions etc. He would be well advised to draw out the block plan to a larger scale such as 1:100 or 1:200 in order that detailed information can be plotted accurately.

It is useful for the builder to cut templates in plastic or cardboard to represent site offices, huts, mixer installations etc. to a scale of 1:100 or 1:200, in order that accurate positions true to scale can be drawn of the builder's installations. By this method, the templates can be moved around the drawing to establish the most suitable locations for equipment.

Another idea is the use of coloured P.V.C. laid on transparent talc over drawings. This has the advantage of staying up when applied but is easily transferred to another part of the drawing.

Care must be taken to superimpose all foundation and drain runs on the layout drawing so that the office accommodation can be positioned well clear of such work and so avoid having to move during the progress of operations.

DISTRIBUTION OF INFORMATION

As soon as possible after the award of the contract, all available information must be circulated to service departments within the builder's own organisation or to specialist Sub-Contractors or Suppliers.

Joinery, such as door frames etc., is often required during the first few weeks of the contract. It is therefore important that joinery details should be sent to the Joiners' Shop to be carefully checked and any queries brought to the builder's attention so that a complete list of details required and queries can be sent to the Architect at an early date.

Often a copy of the Specification or Bill of Quantities is a further guide at this stage and extracts of relevant information should be taken from the Bills and distributed. This not only applies to joinery but also to such items as cast iron drainage, etc.

REQUISITIONS AND INTERNAL ORDERING

At the commencement of all building contracts, numerous small tools and light plant such as shovels, barrows, sledge hammers etc. are required. Every contract seems to follow a similar pattern at this stage with the builder's store or plant yard being inundated with demands for equipment. By referring to experience, the Builder should be able to prepare a 'shopping list' of small tools etc. required at the start of a new contract. This can become a standard form for use on all new work. Provision can be made on the form to indicate the date of delivery; where tools can be grouped together and the requirements for first or second lorry loads can be indicated. By using this simple system, much writing of long lists of equipment can be avoided. An example Check List for starting new contracts is shown as Figure 13.

PROPOSED SITE ORGANISATION STRUCTURE

During the preparation of the estimate, consideration should have been given to the site administrative costs. In Chapter III dealing with Pre-Tender planning techniques, we discussed the preparation of a typical Site Organisation Chart showing the lines of authority and the delegated responsibilities. The appointment of staff who will be directly responsible to the site manager, agent or general foreman should be determined during the planning stages. It is bad practice to launch the agent on the site and leave him to find all his own subordinates.

The introduction of the carpenter foreman into the planning discussions when such items as formwork design are being considered will go a long way to establishing team spirit.

Figure 13.

CHECK LIST AND REQUISITION FOR STARTING NEW CONTRACT

CONTRACT: DATE:

JOB NO.:

LOCATION OF SITE: 1ST LOAD REQUIRED:

GENERAL FOREMAN: 2ND LOAD REQUIRED:

Qty.	Article	Dely.	Qty.	Article	Dely.
	Axes			Picks	
	Barrows			Ranging Rods	
	Battens			Rubber Boots	
	Boning Rods			Safes	
	Brooms			Saw Benches	
	Buckets			Setting out Lines	
	Building Sqrs.			Shovels	
	Chairs			Sign Boards	
	Chisels			Scaffold Tubes	
	Crow Bars			Fittings for ditto	
	Donkey Jackets			Sleepers	
	Electric Fires			Spirit Levels	
	First Aid Boxes			Tapes—Steel	
	Forks			Tarpaulins	
	Hammers—Lump			Tea Urns	
	Hammers—Sledge			Theodolite	
	Hose Pipe 19 mm			Thermometers	
	Hose Pipe Clips			Tool Boxes	
	Hose Repairers			Wash Bowls	
	Jugs, Enamel and Alum.			Water Butts	
	Ladders			Waterproofs	
	Lamps—Danger				
	Lamps—Hurricane				
	Levelling Boards				
	Levels Dumpy				
	Mugs				
	Nail Draw Bars			*Signed:*	
	Notice Boards				
	Petrol Bins				

Further, it will relieve the site manager or general foreman of responsibility for detailed study of such practical problems, leaving his mind free to organise operations to get the contract underway smoothly.

When the site staff has been selected, the anticipated total salaries can be calculated and compared with the estimated allowances in the Bill of Quantities, and thereby maintain an economical and realistic site overhead expenditure.

CONSIDERATION OF OPERATIONAL METHODS

Having spent considerable time on the items of general administration, we will now turn to the operations involved in the construction work. Nothing in the Architect's drawings or the Bill of Quantities or Specification will establish for the Builder which method should be adopted in doing the work. Nor is any indication given whether any job can be done by machine or by hand.

By careful study of the working drawings, together with an accurate knowledge of site conditions (ascertained from the site investigation report) and information about amounts of work from the Bill of Quantities, the contractor must decide the *best* way of performing each operation to be undertaken. If proper tendering techniques are adopted by the company, a Statement of Methods upon which prices have been calculated will be available for consideration by the construction staff on site.

METHODS ANALYSIS

Considerable savings can be made by the Builder if sufficient thought is given to the possibility of a better and more economical methods of carrying out the work than has been priced for by the estimator. The best approach to this is to select several possible methods of executing the major operations, such as excavation, handling concrete, hoisting etc. Each operation will then be priced to discover the estimated cost. These methods and costs will then be compared with the original tender build-up and the most practical and economical method can be selected. Consideration can then be given to the purchase of recommended new plant if it appears from the method analysis to be the best approach to carrying out the work. It must be borne in mind at this stage that although new plant may appear to be the answer to a methods problem, purchase of it may not be in accordance with the Board of Directors' policy which may be to use whatever is available from the firm's own fleet of equipment in order to control capital expenditure.

INVESTIGATING THE SYSTEM OF CONSTRUCTION

During the study of suitable methods of operation, it may occur to the builder that if the design of the building were slightly modified, a better method of carrying out the work could be adopted. Often, constructive suggestions made to the Architect in a discreet manner will be accepted and enable the builder to improve the method of construction, possibly reducing the cost and increase the speed of the contract.

The process of considering alternatives in design can apply equally to the specification for materials. For example, precast units may be replaced by lightweight slabs with a similar load-bearing capacity. This would not affect the appearance of the completed building but would enable the builder to employ his own crane which, due to a limited lifting capacity, would have been useless under the original specification.

THE QUESTIONING TECHNIQUE OF METHODS IMPROVEMENT

It is not intended to dwell at great length upon the subject of Work Study here, but the analytical approach and trained thought process can be of great benefit when making major decisions at the planning stage of a building contract. Once the job has started or a major operation is underway it is often inadvisable to change the plant; and unless the method is improved a bottleneck occurs, delays and unforeseen costs are incurred. How then can we apply Method Study to planning building work? By asking questions which will give us facts about all the operations. Often questions will show that certain work can be avoided or modified. The questioning technique is:

What is being planned?
Who is going to do the job?
When ⎫
Where ⎬ is it going to be done?
How ⎭

71

After each question ask WHY?
  Is the operation necessary?
  Should it be done by another man or men?
  At some other time?
  Elsewhere?
  IN SOME OTHER WAY?

In order to make full use of Method Study during the planning of building operations, it must be appreciated that every situation is a problem and usually has several possible and practical solutions. It must also be realised that the way in which the work was carried out on a previous contract may not be the best solution for this contract. Usually, no two buildings are the same and certainly no two sites are exactly the same. In consequence, the ideal solution for an operation on one contract cannot be exactly the same as that for similar work on another. Because of this a builder must think carefully and find, by method study, the BEST solution for each case.

CHAPTER VI

PRE-CONTRACT PLANNING—PART III

Sub-contractors and Suppliers

'Ten thousand difficulties do not make one doubt . . .'
J. H. Newman
'If you are going to negotiate, then always negotiate from strength' . . . *E. I. Wheatley*

One of the most difficult tasks to be undertaken in the management of building work is the co-ordination and control of specialist services and work carried out by independent firms of Sub-Contractors. Each specialist is usually dominated by the details and problems of his own trade and in consequence is often unable to see how his own particular work or that of the main contractor affects the progress of other specialists.

In these circumstances the smooth running of the contract will be entirely dependent upon the co-ordination and control exercised by the main contractor. We have already considered in Chapter I how controlling and planning are closely connected. To control sub-contracted work or supplier's deliveries, the builder must give himself sufficient time before the job starts to prepare proper plans in conjunction with everyone concerned. By this we mean a carefully considered programme showing how the specialists' work is properly phased into the main contractor's pre-determined sequence of operations.

THE BUILDER'S OWN SUB-CONTRACTORS

Let us first turn our attention to the builder's own Sub-Contractors. At the estimating stage, the builder may have obtained several competitive quotations for each item of work to be sub-let. A summary should now be prepared, listing the quotations in order that a more detailed comparison of Sub-Contractors' tenders can be made.

The following points should be listed on a separate sheet for each item of work for consideration, and it is suggested that a standard pro-forma should be drawn up to contain:

1. Name of Sub-Contractor.
2. Amount of quotation.
3. Terms or discounts.
4. Date of quotation.
5. Any special conditions.
6. Period of time for completion quoted.
7. Remarks.

It may be decided to invite additional firms to submit further quotations. If the builder does decide to extend the invitations to submit prices, this must be done promptly after the award of the contract. Regardless of further 'financial negotiations' that may take place, the construction and planning staff will require the name and particulars of the specialist who will be considered acceptable, subject to the Sub-Contractor's agreement to conform to the builder's programme.

At this stage in the proceedings, it must be firmly established in the minds of everyone associated with the contract that before an order is placed with the Sub-Contractor the builder's construction staff must have the opportunity to discuss and agree a suitable programme of work.

It is bad practice for the technical staff to negotiate prices and place an official order for sub-contract work without references to the construction staff. The latter must then attempt to agree a practical programme after the initiative has been lost because the official order has been placed without tying up the practical approach to the job beforehand. It is a good policy for the builder to make the firm's official order a written confirmation of the verbal agreements made by his own construction and planning staff and the Sub-Contractor's senior management.

By applying this principle firmly to all dealings with Sub-Contractors, the builder will settle all major issues concerning sequence of work, attendances, etc., before the work gets underway. By refraining from placing the order until these negotiations are complete the builder retains the opportunity to go elsewhere if he does not obtain a satisfactory agreement with the Sub-Contractor at the planning stage.

NOMINATED SUB-CONTRACTORS

Control over specialist Sub-Contractors nominated by the Architect or appointed by the client is one of the most difficult problems a builder has to overcome in order to obtain complete co-ordination of all operations involved in the construction programme. It is unusual for the names of the specialists to be available from the Architect at the date when the Contractor's tender is accepted. Usually the Architect has, before inviting tenders, obtained quotations for the specialist work and an amount has accordingly been included as a p.c. sum in the Bill of Quantities.

At the stage when the builder is awarded the contract, the Architect is possibly still undertaking further negotiations or agreeing design problems with these specialists and who will eventually undertake the work is still in doubt. From the builder's point of view this is unsatisfactory because until the specialists are known to him he is unable to formulate a reliable overall programme or master plan for the work. Without some knowledge of the specialist work, he is unable to allow a realistic time in his programme for the Sub-Contractors' operations and, in consequence, will be unable to phase and co-ordinate the work of specialists with his own into a practical programme.

This means that the builder's assessment of the time period for completion of the whole work can be nothing more than a courageous guess.

The builder will then find it almost impossible to control the progress of the contract along reliable predetermined lines. In these circumstances, it is usual for numerous unforeseen problems to occur which cost time and money. These costs must inevitably be borne by the client. It is, therefore, of paramount importance that the builder should obtain at the earliest possible date, a comprehensive list of the names of all specialists to be nominated by the Architect. It is the responsibility of the builder's top management to maintain a firm control at this stage of planning. It may be worthwhile inviting the Architect to attend the builder's pre-contract planning meetings in order to give him a complete picture of the aims and purposes of these techniques and an insight into the need for close collaboration with specialists at this early stage.

The builder must prepare a comprehensive list of all P.C. amounts from the Bill of Quantities and send an immediate request to the Architect for the names of all specialist sub-contractors or suppliers involved.

CO-ORDINATION OF SPECIALIST WORK

Having obtained the names of the various specialist sub-contractors, the builder must arrange to discuss their work in detail. Arrangements can now be made to ask each specialist firm to attend a suitable planning meeting. It can be made clear to the Sub-Contractor that his quotation is under consideration and will probably be accepted subject to a satisfactory agreement being reached between the specialist's representative and the builder's construction staff.

The Sub-Contractor should be made aware that the planning meeting has been arranged to assist the builder to formulate a detailed programme. It is usually a benefit if the Sub-Contractor is requested to bring to the meeting any preliminary drawings he may have in connection with his own work, and a copy of his quotation, particularly if the builder has not yet received the quotation for nominated work from the Architect or Quantity Surveyor. The specialist should also be given an outline of the problems and questions that will be discussed with him at this meeting. A simple questionnaire sheet could be attached to the letter convening the meeting, covering the various points for discussion. This prior notification should enable him to arrive at the meeting fully conversant with the problems and perhaps some of the answers, having had an opportunity of discussing them with his own staff. A typical questionnaire is shown as Figure 14.

After the meeting each questionnaire can be completed and, if satisfactory agreement has been obtained with the specialist, the details of the arrangements made can be forwarded to the technical staff to incorporate the various points in the builder's official order for the work.

ORGANISING SPECIALIST SUPPLIERS

With many Quantity Surveyors and Architects it is normal practice to insert in the Bill of Quantities such clauses as:

'Reconstructed Artificial Stone for copings is to be in

Figure 14.

SUB-CONTRACTORS QUESTIONNAIRE SHEET

CONTRACT: DATE:
JOB NO.:
FIRM: REP.:   TEL. NO.:

Question	Answer
1. How long will it take to complete the whole works?	
2. What is the proposed labour strength?	
3. What is the proposed sequence of work?	
4. How many visits will have to be made to the site?	
5. What information is required prior to commencement of work on site? (drawings etc.)	
6. What is the minimum notice which may be given for commencement?	
7. How soon can the main contractor be furnished with information concerning size and positions of holes and chases?	
8. At what stage will it be possible to take any site dimensions required?	
9. What drawings or information is required from *other* specialists?	
10. What information is required from the Main Contractor?	
11. What type of storage space is required and when?	
12. What attendances are required?	

77

accordance with samples submitted to and approved by the Architect.'

The builder should carefully read through the Bill and make an abstract of all items of this nature, carefully listing all materials for which samples are required. Specialist suppliers must then be found and samples of material sent to the Architect for approval. It is wise to collect all these items together at the start of the contract so that plenty of time will be available for the Architect to reach a final decision. Care must be taken to obtain firm quotations at this stage, together with reliable periods of delivery.

In many cases the Architect will specifically nominate firms as suppliers and send the builder the quotations to be accepted and orders to be placed. It is advisable for the builder to write to these suppliers before placing firm orders and to obtain delivery dates in writing. Often working drawings have to be prepared and the builder must find out how long it will take to manufacture and deliver the material *after* being in possession of finalised working details.

This is a regular occurrence and in consequence a 'standard' letter can be prepared for this purpose. A specimen typical letter to suppliers is shown as Figure 15.

MATERIAL DELIVERY SCHEDULES

A programme, however accurately prepared, cannot be carried to a successful conclusion unless arrangements have been made to get materials etc. to the site as and when required.

A systematic analysis of the Bill of Quantities must be undertaken and comprehensive lists prepared of material required showing the amount and anticipated date of delivery. In the case of materials such as bricks, where many deliveries are required, the date of the first consignment required should be stated.

Figure 15.

Messrs. Reconstructed Stone Ltd.,
Jacobean Lane,
West Drayton.   Date:

Dear Sirs,
    Messrs. Brown & Co., Proposed Factory, Bradford.
We have been advised that we have been successful in obtaining

78

the contract mentioned above and we are now considering your quotation for stone copings.

In order to assist us in preparing a detailed programme for the contract, will you please advise a firm delivery date or state the period required from receipt of final approved detailed drawings before delivery can be quaranteed. It would also be of assistance to us if you can outline what drawings you will require in order to fabricate this material and ensure prompt delivery.

Yours faithfully

When the materials schedule is complete, the builder can refer to the replies received from the 'letter to suppliers' mentioned earlier. The date of delivery guaranteed by the suppliers against each item can be compared with the builder's planned requirements. Any anticipated delivery problems can be foreseen in this way and arrangements made to advise the Architect accordingly. In these circumstances a change in the specification will often enable the builder to order material that will be available within the required period for delivery and thereby avoid unnecessary delay in the future.

By referring again to the supplier's letter the builder can prepare a comprehensive list of detailed information required by the various suppliers and the deadline dates when these details must be available. This enables the Architect to prepare his own list of priorities and arrange continuity of information for the contract and his suppliers throughout the contract. A typical example of material delivery schedule is shown as Figure 16.

Where the builder is dealing with nominated Sub-Contractors or Suppliers, he must retain the initiative during the negotiation stages, particularly before orders are placed. During the finishing stages these specialists are the controlling force behind the progress of work. Delays caused by Sub-Contractors and Suppliers involve the contractor in considerable loss and expense which was probably not foreseen at the estimating stage. This situation can often be improved if the builder clearly understands that although the Architect has nominated the firm as a specialist, the placing of the official order is entirely dependent upon the main contractor making satisfactory arrangements with the specialists and obtaining their

Figure 16.

MATERIALS DELIVERY SCHEDULE

CONTRACT: CONTRACT NO.:   SHEET NO.:

Order or Requisition	Material	Supplier	First date of delivery	Total Quantities	Remarks

Signed:

80

complete agreement to conform to the programme and delivery dates.

No advantage whatsoever can be gained by the builder placing the order and then entering upon negotiations with the firm concerned. Should difficulty be encountered in obtaining complete agreement at this stage, it should be suggested to the Architect that an alternative specialist is selected. It cannot be over-emphasised that a builder's official order, whether to Sub-Contractor or Supplier, is not only a copy of the quotation outlining the financial arrangements but must always be the written confirmation of verbal agreement made between the contractors' construction staff and the specialists' senior and responsible representatives.

F

CHAPTER VII

PRE-CONTRACT PLANNING—PART IV

programme calculations

and charts

'A Manager who tries to run his business without accurate information and without the information contained in charts is like the Surgeon trying to perform an operation without hands or a Violinist trying to play without his violin . . .'   *Sir Graham Cunningham, K.B.E.*

THE VALUE OF PROGRAMME CHARTS

In the past builders have not been schooled in graphical methods of presenting facts, since such methods were little used by management until recently. Many have got along very well without them up to the present time and have yet to learn that any advantage can be gained from their use. Many are firmly convinced that there is something complicated and mathematical about them and being no mathematicians themselves, and possessing also the practical man's mild contempt for the sciences, object to the use of charts in business.

'. . .[1] One more reason for this unpopularity may perhaps be added, and that is when a chart is properly made and therefore instructive, it forces the reader to think. Most people object strongly to thinking—or at least to thinking hard. They are quite prepared to do a quick calculation on the back of an envelope, quote a price and take an order, but to sit down to a set of interlocked business charts and think hard and collectedly as to what lessons can be gleaned from the whole series of results is a task that calls for will power and concentration and, therefore, will never be popular . . .'

We have considered in previous chapters how management in building is based on control obtained through comparison. In order to control building work the builder must be able

T. G. Rose. *Business Charts.* Pitman 1959, p.2.

to make certain that the progress of work at the moment compares favourably with what had been planned.

It is obvious therefore that a system of charts must be devised to set out the information that will be required in the clearest and simplest form.

In order that full value is obtained from charting the information, the builder will need to know the following:

1. *Commencement date of Key Operations*

The major operations or 'key' operations which have a controlling effect upon the construction sequence must be clearly shown with agreed dates for commencement.

Such work as structural steel framework, roof covering, etc. are important 'mile stones' in the progress of the work and, generally speaking, accurate charting of these operations will ensure that the programme of work is feasible and practical.

2. *Period of time for Key Operations*

Accurately calculated time periods for key operations are essential. Approximate and vague assessments leave a great margin for error. Key operations badly planned will cause the programme to fall rapidly out of date and thereby serve no useful purpose in controlling progress.

3. *Phasing and Sequence of Operations*

Careful consideration of the start of each operation in relation to other operations will ensure practical sequences. Notes on the chart explaining the reasons for the selected sequence of major items will clarify the other information on the chart.

4. *Planned Labour Strengths*

In addition to the period of time for the work, it is important to show the number of operatives and the balance of gangs required to achieve the planned completion dates.

5. *Planned Plant Requirements*

The type of plant to be used and its anticipated output, required to achieve the programme date must be shown. Additional information may be calculated from the programme but this will be discussed later on.

METHODS OF CALCULATING THE TIME PERIOD FOR OPERATIONS

In many cases the contractor will have to rely upon the accuracy of the Bill of Quantities to form the basis for calculations. If insufficient detail is available from the Bill of Quantities or if the amounts are provisional and subject to re-measurement, the builder will have to carry out accurate measurement from working drawings.

PROGRAMMED OPERATIONS

Before information can be abstracted from the Bill of Quantities, an accurate list of building operations must be prepared. This is often called the List of Programme Elements and is usually broken down for greater detail than simple 'trade order'. Usually the sequence of work outlined in the Standard Method of Measurement for Building Work as used by the Quantity Surveyor when 'taking off' is far too general for this purpose.

In other words, the builder will have to break down an operation into several programme elements. For example, the operation of 'Concrete to columns not exceeding $0 \cdot 10$ m^2 sectional area' would possibly be analysed into:
1. Concrete to column-starter at ground floor level.
2. Concrete to columns ground floor to underside first floor beams.
3. Concrete to column starter—first floor level.
4. Concrete columns—first floor—roof.
   etc.

CALCULATIONS BASED UPON ACTUAL MEASUREMENT

After the list of Programme Elements has been compiled the 'work content' for each item must be obtained. Each operation, or element, can be now considered separately and the work measured from detailed drawings—preferably finalised working drawings.

The quantity of work involved for each element will then be available. These quantities can be multiplied by either:
 (i) Nett labour rates.
 (ii) Output rates.

In Chapter IV we considered the pricing of the Bill of Quantities into sub-divided unit rates. The nett labour or

Figure 17.

PROGRAMME ELEMENT
CALCULATION SHEET

CONTRACT _____ J.S. DOBSON

CONTRACT N°. 1374

ELEMENT N°. EXC. 4.

B/Q REF.	QTY.	UNIT MEAS	NETT RATE LAB.	NETT RATE PLT.	TOTAL LAB.		TOTAL PLANT		TOTAL PRICE		NOTES
$4/4^B$	180	m^3	$2\frac{1}{2}p$	7p	£5	90	£16	52	£22	42	Serious consideration must be given
$4/4^H$	40	m^3	3p	$7\frac{1}{2}p$	£1	56	£3	90	£5	72	to use of 955 Traxcavator. Suggest
$4/4^J$	10	m^3	3p	$9\frac{1}{2}p$	£0	30	£1	14	£1	56	we plan on this.

£29 | 70 TOTAL.

PROPOSED PLANT

955 EXCAVATOR.

£21 56 ÷ £4.50 (M/C RATE) ___ = 5.00 ___ PLANT
PER HR ___ HRS

∴ ___ Trax. ___ (M/C FOR) ___ 5 ___ HRS

AT £4.50 (HIRE RATE) COSTS £22.50
PER HR

PROPOSED LAB. STRENGTH

ONE BANKSMAN.

£7 52

TOTAL NETT LABOUR

X ___ 2.25 ___ ÷ ___ 44 ___ = ___ 4¼/11 ___ MAN WEEKS
(£0.45 per hour)

DESCRIPTION OF OPERATION

Excavate to reduce levels not exceeding 1.50 m deep

plant rates associated with the relevant operation are taken from the Bill of Quantities and is multiplied by the measured quantities and the total labour value established for the item.

This labour value divided by the current hourly rate used by the estimator at the time of pricing will produce the total man hours for the operation. If no nett labour or plant rates are available from the Bill of Quantities, the actual output rates used by the estimator for pricing purposes must be obtained. By simple multiplication the total man hours can be determined.

It is important throughout these calculations that the estimater's unit costs are used as a basis for calculations. At a later date *actual* costs can be produced and a proper comparison will reveal if inaccurate output rates are being used by the company for estimating purposes.

CALCULATION BASED UPON INFORMATION AVAILABLE IN THE BILL OF QUANTITIES

An alternative method of calculating the work content from the List of Programme Elements is to use the Bill of Quantities. It must be firmly established that the Quantities contained in the Bill are accurate and have been measured from reliable working details.

If approximate quantities are used the eventual programme will be approximate and in consequence, when work commences, slight deviations in the amounts usually cause the programme to become impractical.

Having decided to use the Bill of Quantities, each item in the Bill can be abstracted under the relevant heading for each Programme Element. It is useful to have a simple pro-forma sheet for this purpose. An example is shown as Figure 17.

The nett labour contained in the estimate for each Programme Element is obtained by multiplying each relevant item in the Bill by the priced labour values and obtaining the total. The total money value for each element from these calculation sheets should be converted into Total Man Weeks. This can be achieved by a simple formula:

$$\frac{\pounds \times K}{40}$$

Figure 18.

SEQUENCE STUDY CHART

CONTRACT J. S. DOBSON LTD.　　**CONTRACT Nº.** 1374.　　**DATE** 30.9.70.

AMNT.	UNIT	Nº.	DESCRIPTION	METHOD	STD.	HRS.
21	m²	1	FORMWORK TO COLS.	PLYWOOD & CLAMPS	2.15	45.00
1.81	m³	2	CONCRETE TO COLS.	LIGHTWEIGHT HOIST	8.47	15.00
52	m³	3	EXCAVATION GND. BMS.	HAND	2.92	152.00
78	m²	4	BLINDING GND. BMS.	10/7 MIXER & BARROWS	.36	28.00
65	m²	5	FORMWORK : :	PLYWOOD	1.29	84.00
500	kg	6	REINFORCEMENT :	FIX ONLY – SUBLET	.04	20.00
22	m³	7	CONCRETE : :	10/7 MIXER & BARROWS	4.25	93.50
69 / 530	m³ / m²	8	HARDCORE BACKFILL	HAND	1.30 / .42	90.00 / 222.00
		9	SCAFFOLDING	SUBLET		
270	m²	10	BRICKWORK GND - 1ST	PACKAGED BRICKS	2.40	648.00
304	m²	11	SCAFFOLDING	SUBLET		
97	m²	12	FORMWORK TO COLS.	PLYWOOD & CLAMPS	1.62	157.00
8.83	m³	13	CONCRETE TO COLS.	LIGHTWEIGHT HOIST	8.47	75.00

Column headers: QUANTITIES (AMNT. | UNIT) — OPERATION (Nº. | DESCRIPTION) — METHOD — OUTPUT (STD. | HRS.) — followed by weekly day columns M T W T F S S repeated.

For this purpose K represents the result when the average basic hourly rate—a compromise between craftsmen and labourers rate—is divided into a pound. For example, if the average hourly rate between craftsman and labourer was £0·45 then K would be £1·00 divided by £0·45 which is 2·25.

The figure 40 is used as an arbitrary figure being about 10% less than the average working week of say 44 hours. The 10% adjustment is to allow for inclement weather and will of course be subject to adjustment depending upon the location of the site.

It must be admitted that this formula for calculating the man weeks is not entirely correct. Although taking an *average* hourly rate does produce a slightly inaccurate picture it does serve as a simple rule of thumb method which should be sufficiently acceptable for use at the overall planning stage.

SEQUENCE CHARTS

Having carried out the calculations required to determine the work content for each operation in the list of Programme Elements the information must now be plotted on a chart in order to obtain a much clearer picture of the anticipated progress of the work. An example of a Sequence Study is seen in Figure 18.

All operations will be shown carefully related to one another and a realistic and practical phasing of the sequence of work clearly shown.

The most important factor to bear in mind during the preparation of this type of chart is the practical commencement date for each operation. Dependent upon this will be the required delivery dates, commencement dates etc.

The overlapping of trade operations must be carefully considered at all times and the person responsible for preparing the chart must make sure that if two operations are shown to be carried out simultaneously, adequate working space is available and that the arrangement is a practical proposition.

THE MASTER PLAN

This should be a simple but comprehensive picture of the contract as a whole, embracing the information calculated from detailed sequence charts and showing the phasing of all

Figure 19.

OPERATION – DESCRIPTION	PLANT	SUB CONTRCTR.
1 PRELIMINARIES – Erection Site Offices.	Standard Huts	Regan Bros.
2 DEMOLITION – Structure.		G. P. O.
3 SITE SERVICES.		Haddock.
4 EXCAVATION – Reduced Level.	CASE 1150	
5 DRAINAGE WORK.	57R.5	
6 EXCAVATION – Stanchion Bases.	JCB AC DIGGER	
7 CONCRETE – Stanchion Bases.		
8 FILLING. – Hardcore Backfilling.	Vibro Rose	
9 CONCRETE – Columns and Beams.	Coleman Pourer	
10 EXCAVATION – Foundation Beams.		
11 CONCRETE – Foundation Beams.		
12 CONCRETE – Suspended Floors.		
13 STAIRCASES.	Mortar Pan.	Jacksons Ltd.
14 BRICKWORK – Superstructure.		
15 CARPENTRY WORK – First Fixing.		Wilfred Robbins
16 CRANE RAIL INSTALLATION.		Insulated Roole Ltd.
17 ROOF COVERING.		Sherbourne Ltd.
18 STRUCTURAL STEEL WORK.		W. Farrer
19 PLUMBING WORK.		S. Warner
20 ROOF LIGHTS.		Jas. Gibbons
21 CURTAIN WALLING.		R. M. Douglas.
22 ROOF COVERING – Asphalte.		J. Gibbons
23 ROOF LIGHTS – Member.		Williams Bros.
24 HEATING INSTALLATION.		
25 ELECTRICAL INSTALLATION.		Mather & Platt
26 SPRINKLER SYSTEM.		J. Gibbons
27 JOINERY WORK – Doors, Skirtings, etc.		Mather & Platt
28 SHUTTERS & DOORS.		G. Reynolds (William De Dozen)
29 PLASTERING WORK.		Hill Bros. Ltd.
30 GLAZING WORK.		Sherbourne.
31 METALWORK.		Fisson-Plywood.
32 W. C. PARTITIONS.		Acoustic Cabling.
33 JOINERY FINISHINGS.		
34 FLOOR FINISHINGS.		
35 CONCRETE – Ground Floor.		Johnson Firg.
36 GRANOLITHIC PAVINGS.		
37 DECORATIONS.		Raysham Ltd.
38 FLOOR COVERINGS.		Raysham Ltd.
39 VENITIAN BLINDS.		

REQUIREMT. SYMBOLS
◇ DETAILS
← NOMINATIONS
▽ REINFORCEMENT
○ SAMPLES
● SCHEDULES

REMARKS

key operations. It illustrates graphically how the sequence of work will have to be arranged to achieve a completion for the work within the contract period.

The extent to which information should be shown on the Master Plan will depend upon the size and complexity of the contract.

As the chart is intended to show the solutions to problems foreseen at the planning stage of the job, it is logical that as no two contracts are the same, no two solutions will be the same. Therefore, no two charts can be indentical and the standardisation of Master Plan Charts would be difficult. However, a typical illustration of an overall programme chart, being the Master Plan for a hypothetical contract, is shown as Figure 19.

It will be noticed that provision has been made to record actual progress on this chart. This form is frequently referred to as the Overall Programme Chart.

This type of chart is widely used and is extremely effective as it shows not only the sequence of work but also the labour and plant requirements to achieve the planned output of work.

A check list to ensure that the chart contains sufficient information to assist the management in the smooth running of the job, is shown below.

Does the chart show?

1. Commencement date for each operation.
2. Period of each operation.
3. Sequence of operations.
4. Phasing of operations.
5. Number of operatives required to carry out each operation.
6. Mechanical Plant required to carry out each operation.
7. Delivery date of material for each operation.
8. Date when drawings and information must be available to carry out each operation.
9. Names of Sub-contractors involved.
10. Names of specialist suppliers involved.
11. Holiday periods.
12. Provision to record progress.

COLOUR CODES FOR CHARTS

In order to identify operations carried out by similar trades, and to show a proper continuity of work for an individual trade, colour can be introduced for work to be undertaken by separate trades.

A suggested colour coding for charts is shown below:

Excavation work—Brown.
Concrete work—Green
Formwork—Orange
Steelwork—Blue
Brickwork—Red
Joinery—Yellow
Carpentry—Yellow
Pavior—Grey
Masonry—Light Green
Sub-contracted work—Purple
Plumbing—Dark Blue
Decorations—Pink

This can, of course, be elaborated to suit the builder's requirements.

REQUIREMENT SCHEDULES

Having produced the Master Plan as an overall picture of the project, further information can be extracted from the chart by careful study and an analytical approach to the problem of 'What must we get to the site in order to achieve this programme'.

The progress of work will be entirely dependent upon the control of resources. Unless every effort is made to get material, men and sub-contract services on the job at the right time, the expense of planning before commencement of work on site will be wasted.

In order to control these resources, schedules of requirement must be prepared showing what is wanted and when it must be available. This information should be readily available from the programme chart.

LABOUR SCHEDULE

Planned labour requirements for each operation must be

summarised to show the number of skilled and unskilled men required each week. This may be shown in the form of a simple graph for each trade or by simple addition in every trade at the bottom of the Master Plan for each week of the contract. This gives the Contractor an advance warning of his requirements for such men as reinforcement steel fixers, etc.

PLANT SCHEDULE

A list of Mechanical Plant giving the dates when required and an approximate indication of the period of hire will assist the builder to make arrangements well in advance to secure equipment not owned by the Contractor himself.

MATERIAL SCHEDULE

The Material Schedule has already been explained in Chapter III but the dates when the first delivery will be required can be extracted from the Master Plan and orders placed well in advance.

DETAILS AND DRAWINGS SCHEDULE

No Architect can foresee the order and sequence in which information will be required by the builder. A schedule listing all outstanding details, information etc. can be prepared and dead-line dates furnished from the Master Plan. This document will often prove to be of great value to the Builder in the settlement of claims. To be able to let the Architect have an exhaustive list of outstanding information, complete with dates when required, puts the contractor in an advantageous position if the work is held up due to lack of such details.

SUB-CONTRACTORS' SCHEDULE

Lists of all Specialists, together with planned commencement dates and periods of completion, will help the technical staff to place accurate and reliable orders with all Sub-Contractors.

OTHER SCHEDULES

Various other lists of requirements can be produced from information provided by the Master Plan, such as scaffolding, formwork material, etc. The extent to which the builder will take the preparation of these schedules will depend upon the size of the contract and above all the time available.

For the builder to obtain full benefit from these charts and requirement schedules, they must be kept up to date throughout the running of the job. Actual progress and actual deliveries must be recorded, in order that a proper control over the contract and the material supply sources can be maintained.

CHAPTER VIII

SHORT TERM PLANNING

'. . . Executives must exercise care to keep their plans and systems alive, vital and responsive to changing conditions . . .'

In Chapter IV we have already considered the preparation of the Master Overall Plan before starting on site; this provides the supervision and the management with a broad view of the Contract as a whole. In the definition and purpose of planning it was explained that the effectiveness of planning is directly proportional to the extent to which the component which is being planned can be controlled. For the benefit of overall planning to be maintained after work has commenced, the progress must therefore be controlled and the work ahead kept under constant review.

SHORT TERM PLANNING DEFINED

Short term planning is a periodic consideration of the progress to date for each and every operation in the Master Overall Plan. It also embraces a 'Forward View' of the work to be carried out for a short period ahead. This period will depend upon the circumstances but it usually varies between 3 to 5 weeks. The aims of Short Term Planning are illustrated by a simple diagram as Figure 20.

THE PURPOSE OF SHORT TERM PLANNING

The purpose of preparing short term plans is to keep the Master Plan 'alive' and responsive in the light of changing or unforeseen circumstances.

Nothing can be gained by leaving an out of date and unrealistic plan pinned up in the site office. This document must be kept up to date. Not only must it show the progress of

Figure 20.

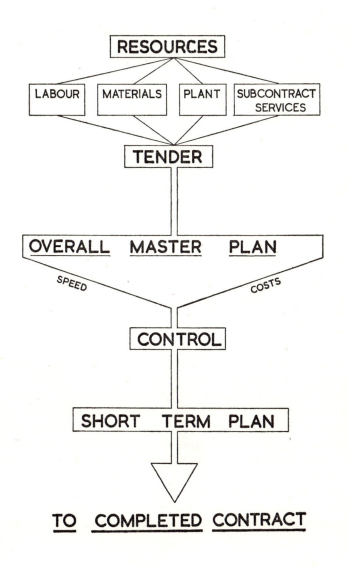

AIMS OF

SHORT TERM PLANNING

work, but also provide an accurate forecast of operations to be undertaken in the immediate future.

While accurately recording progress and considering the future, every effort should be made to recover any 'lost time' and the aim should always be to get back on to the original programme, if possible. Should the contract be running to schedule, or possibly ahead, a great benefit can be obtained by reviewing future delivery dates etc. for any items that are still outstanding.

Each operation to be carried out during the short period ahead should be carefully considered and final decisions made to ascertain:

(a) The quickest method.
(b) The most economical method. $\left.\right\}$ Not always the same.
(c) The sequence of work.

THE FIVE-WEEK SHORT STAGE PLAN

This is a typical example of a simple type of Short Term Plan and is used to review progress and establish a 'Forward View' over five consecutive weeks. It is often referred to as The Monthly Plan. An example of a Five Week Plan is attached as Figure 21.

SEQUENCE OF OPERATIONS

The work in progress should be carefully studied in relation to the overall Master Plan. All operations *behind* programme to date and all operations *in progress* to date, should be carefully listed. A 'forward view' of work originally planned to be carried out, as laid down in the Overall Master Plan, over the period, should then be considered. Every operation involved must be carefully analysed and any new operation added to the list of operations for work behind or in progress. A complete summary is then available, showing every item of work that will have to be completed in order to bring the work on site up to programme in accordance with the original Master Overall Plan.

This is a very important part of the technique. Unless a complete and comprehensive list of operations is prepared at this stage, the Short Term Plan will be inaccurate. Within a few days so many items of work will occur that were not programmed that the whole Short Term Plan becomes imprac-

Figure 21.

G

FIVE WEEKLY PROGRAMME OF WORK

CONTRACT J. S. DOBSON LIMITED CONTRACT No. 1374 PERIOD APRIL 1st TO APRIL 30th

QUANTITIES AMNT.	UNIT	OUTPUT STANDRD.	LABOUR TOTAL HRS.	GANG SIZE	GANG HRS.	GANG DAYS	PLANT TYPE & NUMBER	HOURS	DAYS	OPERATIONS No.	DESCRIPTION	TRADE OR SUB-CONTRS.
80	m^3	2.925	234	6L		5 D.				1	EXCAVATION. Stanchion Bases.	
80	m^2	0.36	29	4L	7 HRS		DUMPER			2	BLINDING. Stanchion Bases.	
1116	kg	0.04	44	2S.F.		3 D.				3	REINFORCEMENT. Stanchion Bases	
25 Bases		2.00	50	2C		3 D.				4	FORMWORK. Stanchion Bases.	
33	m^3	4.03	133	5L		3½ D.				5	CONCRETE. Stanchion Bases.	
3248	kg	0.04	129	2SF		8 D.				6	REINFORCEMENT. Columns.	
20.8	m^2	2.15	45	2C		3 D.				7	FORMWORK. Columns (G.F.L).	
1.81	m^3	8.48	15	4L	4 HRS					8	CONCRETE. Columns (G.F.L).	
52 Add 60%	m^3	2.925	152 91	4L		7½ D.		245 HRS		9	EXCAVATION. Ground Beams.	
										10	HARDCORE Backfilling Columns (Internal).	
78	m^2	0.36	28	4L	7 HRS					11	BLINDING. Ground Beams.	
65 Add 100%	m^2	1.29	84 84	4C		5 D.		168 HRS		12	FORMWORK. Ground Beams.	
500	kg	0.04	20	1SF		2½ D.				13	REINFORCEMENT Ground Beams.	
22	m^3	4.25	93.50	5L		2½ D.				14	CONCRETE. Ground Beams.	
69 530	m^3 m^2	1.30 0.42	90 222	2L 5L	37 HRS	5½ D.				15	HARDCORE Backfilling Ground Beams.	
										16	SCAFFOLDING Working (Internal)	
270	m^2	2.40	648	4B		20½ D.				17	BRICKWORK. G.F. - 1st Fl.	
304	m^2	0.30	91	2SC.		5½ D.				18	SCAFFOLDING for brwk.	
97	m^2	1.62	157	4C		5 D.				19	FORMWORK. Cols. G.Fl - 1st Fl.	
8.83	m^3	8.47	75	4L		2½ D.				20	CONCRETE. Cols. G.Fl - 1st Fl.	
220	m^2	0.30	66	2SC		4 D.				21	SCAFFOLDING. Working (External)	Kwickform
135	m^2	3.25	437	4C		13½ D.				22	FORMWORK. 1st Fl. Beams.	
335	m^2	0.54	180	4C		6 D.				23	FORMWORK. Suspended 1st Fl.	

APRIL. Weeks 7, 8, 9, 10, 11 — M T W T F S S

GENERAL NOTES

REQUIREMENT SYMBOLS

◊ DETAILS
← NOMINATIONS
◁ REINFORCEMENT
○ SAMPLES
● SCHEDULES

BRIEF SCHEDULE OF REQUIREMENTS

A	STEEL REINFORCEMENT	ORDER NO. 2771	✓
B	PRE-MADE COLUMN SHUTTER	REQ. NOTE. 0074	✓
C	HARDCORE	ORDER NO. 2784	✓
D	SCAFFOLDING	REQ. NOTE. 0101	✓
E	BRICKS - COMMONS	ORDER NO. 2770	✓

ticable and, above all, is no longer a suitable and reliable document to control the progress of the work.

PRE-MEASUREMENT OF THE WORK IN THE SHORT TERM PLAN

Each operation included in the list of descriptions of work under consideration in the 'forward view' must be assessed to obtain a realistic labour force and period of completion for carrying out the work. Every 'description of work' should now be broken down into smaller items or operations and the proposed work measured in accordance with the requirements. Work values or standard output values already supplied at the Overall Planning Stage are then applied to the measured quantities and the total man hours, or days, established for each operation. Where output standards are not available, they must be furnished by the Head Office and for this purpose the Agent will need to maintain contact with the Estimator.

The Agent or General Foreman then decides upon the size of labour force or gang required to carry out the operation economically and to obtain maximum practical efficiency. At this stage, a simple but effective questioning technique might be applied to great advantage:

How is this operation to be carried out? Why?

How else could this operation be carried out?

Who is carrying out this operation? Why?

Who else could carry out this operation?

When is this operation being carried out? Why?

When else could this operation be carried out?

Where is this operation being carried out? Why?

Where else could this operation be carried out?

Many of these questions may not apply to every operation, but the application of this technique will ensure that the operation has been considered from every aspect before a final decision is reached.

With an agreed labour or gang strength established, a simple division of the total man hours or days allocated for the operation will provide the following information:

    (i) Number of skilled and unskilled operatives required to carry out the operation.

(ii) The period of time required by the operatives collectively to perform the work involved in the operation.

PHASING OF THE SEQUENCE OF WORK

All operations must now be carefully related to one another to obtain a realistic and accurate phasing and sequence of the work. The practical commencement must be established for each operation, the overlapping of trades and operations being at all times carefully considered. Work to be carried out at infrequent intervals and intermittently with other operations must be carefully phased and plotted to establish planned commencement dates for dependent operations.

SCHEDULE OF REQUIREMENTS

The complete Short Term Planning Chart, properly prepared, is not the sole product of this technique. Unless considerable thought is given to the requirements of the site well in advance, the continuity of work, however well planned it is, will still be disrupted. When a detailed sequence of operations has been prepared and accurately calculated as a practical 'forward view', it is essential to check that everything is already ordered for the five weeks being considered and to ensure that delivery will take place at the required time. If there are variations to the contract, fresh ordering may be necessary. If so, this must be done well in advance. The following procedure, when adopted, will ensure that all requirements for the Short Term Plan are anticipated well ahead.

Each operation listed in the Short Term Plan is now considered separately. The following questioning techniques is then applied:

What material is required? How much? When? Where?

What plant is required? How much? What size? What type? When? Where?

What labour is required? What trade? How many? When?

What details are required? From whom? What is the deadline date?

What Sub-Contractors are required? When to start? Where?

Suitable forms of Requirement Schedules are attached as Figures 22 and 23.

Figure 22.

SUB-CONTRACTORS REQUIREMENT SCHEDULE

CONTRACT J. S. DOBSON LTD.   CONTRACT No. 1374.   DATE   25th August, 1970

No.	DESCRIPTION	SUB CONTRACTOR OR SUPPLIER	PERIOD REQUIRED BEFORE COMM'MNT AFTER ORDER	PLANNED COMMENCEMENT DATE	DATE WHEN ORDER MUST BE PLACED	DEADLINE DATE FOR ARCHITECTS INSTRUCTION
1.	STEEL REINFORCEMENT.	REINFORCEMENT LTD.	3 - 4 WEEKS.	14.9.70	IN HAND (ARCHITECT VERBAL)	------
2.	PAT. GLAZING MONITOR ROOF LIGHTS.		9 WEEKS.	26.10.70	AT ONCE	AT ONCE
3.	ROOF AND WALL CLADDING.		8 WEEKS.	19.10.70	AT ONCE	AT ONCE
4.	FELT ROOFING.		3 WEEKS.	26.10.70	25.9.70	21.9.70
5.	STEEL FRAME.	STEELWORK LTD.	IN HAND.	5.10.70	IN HAND (ARCHITECT VERBAL)	------
6.	IRONMONGERY.		8 - 10 WEEKS.	16.11.70	2.9.70	31.8.70
7.	METAL WINDOWS.		3 WEEKS.	19.10.70	18.9.70	14.9.70
8.	DOMELIGHTS.		9 WEEKS.	9.11.70	1.9.70	28.8.70
9.	ROLLER SHUTTERS.		6 - 8 WEEKS.	30.11.70	9.10.70	5.10.70
10.	W.C. PARTITIONS.		ORDER SITE SIZES 9 - 10 WEEKS.	7.12.70	18.9.70	7.9.70
11.	SANITARY FITTINGS.		6 - 8 WEEKS.	30.11.70	2.10.70	28.9.70
12.	HOT WATER INSTALLATION.		6 WEEKS.	9.11.70	25.9.70	21.9.70
13.	GRANOLITHIC PAVING.		3 WEEKS.	16.11.70	16.10.70	12.10.70

Figure 23.

REQUIREMENTS SCHEDULE OF DRAWINGS & DETAILS

CONTRACT ___ J. S. DOBSON LTD. ___ CONTRACT Nº. ___ 1374. ___ DATE ___ 30.7.70 ___

REF.	DESCRIPTION OF INFORMATION REQUIRED	DEADLINE DATE	REMARKS
1.	POSITION OF TEMPORARY SCREEN.	14.9.70.	
2.	DETAILS OF REINFORCEMENT TO STRIP FOUNDATIONS.	14.9.70.	
3.	DETAILS OF CONCRETE STEPS.	21.9.70.	
4.	DETAILS OF PRECAST CILL AND COPINGS.	7.9.70.	
5.	DETAILS OF RAGBOLTS IN CONCRETE BEAM.	21.9.70.	
6.	DETAILS OF JOINERY.	21.9.70.	
7.	DETAILS OF FLOOR CHANNEL.	14.9.70.	
8.	DETAILS OF LAVATORY BLOCK.	7.9.70.	
9.	DETAILS OF COMBINED LINTOL AND CAVITY TRAY.	21.9.70.	
10.	DETAILS OF DOME LIGHTS.	27.10.70.	
11.	DETAILS OF ROADS AND FOOTPATHS.	26.10.70.	
12.	DETAILS OF EXPANSION JOINTS AND DOWELS IN FLOOR.	28.9.70.	
13.	DRAWINGS OF DRAINAGE LAYOUT AND M.H. INVERTS.	7.9.70.	
14.	HEATING SCHEME LAYOUT.	14.9.70.	
15.	COLOUR SCHEME AND FINISHING SCHEDULE.	16.11.70.	

SHORT TERM PLANNING AND THE SITE FOREMAN

The Agent or General Foreman must be held personally responsible for formulating the Short Term Plan. We considered in Chapter IV how, before commencement of work on site, the Site Agent or General Foreman in charge of the contract was responsible, in conjuction with the rest of the Planning Team, for the preparation of the Master Overall Plan. In order to make the Plan effective, he will need to keep his progress under observation. A proper control must therefore be kept on site by him and any necessary action be taken well in advance to ensure the continuity of work as laid down in the Overall Master Plan.

In addition, he must be responsible for the complete preparation of a schedule of outstanding requirements. This should then be submitted to his Contracts Supervisor or Manager for prompt attention.

In order that the Contracts Management is kept informed of detailed progress from time to time, it is also essential that the Contracts Manager or Contracts Supervisor should participate in the preparation of the Short Term Plan. The Manager or Supervisor is then better enabled to keep a broad control over the progress of work and provide advice and assistance to the Agent or General Foreman when most needed.

THE MANAGEMENT AND SHORT TERM PLANNING

Adequate training and proper supervision must be provided from Head Office whenever it is considered necessary. In addition, there must also be an executive at Head Office who is responsible to the Management for seeing that the Short Term Plan is prepared at frequent intervals, as agreed, or more frequently if necessary. As already explained, the Contract Supervisor or Manager should attend the site for the preparation of the Short Term Plan in order to advise and guide the Agent and or General Foreman in order that a practical and realistic 'Forward View' of the work in hand is obtained.

When schedules of outstanding requirements are prepared at this stage, it is the responsibility of the Contract Supervisor or Manager to see that all items outstanding in the way of materials, plant, labour, details, etc. are obtained and despatched to the site without delay.

THE BUILDER'S QUANTITY SURVEYOR AND SHORT TERM PLANNING

The Contract Surveyor should be availabe, if required, to provide quantities by measurement for any item or operation included in the Short Term Programme.

The Surveyor should also be available to measure the extent of work already completed (if required by the Agent and or General Foreman) in order to determine accurately the extent of the progress of work up to date prior to the preparation of the Short Term Programme. It is possible that during the preparation of the monthly valuations and applications for monthly payments, the Builder's Surveyor will carry out measurements of work which has been completed. This will often provide valuable information for Short Term Planning. The Surveyor's monthly applications for certificate payments are closely allied to progress measurement to assess the effectiveness of site control in relation to the Short Term Plan. It is possible, in many cases, for the Builder to devise a system on certain types of work which will reduce the amount of duplication of measuring which would no doubt occur unless the system is closely integrated.

In the larger building firms it is an advantage for a member of the Head Office staff to be available to guide and instruct the Agent or General Foreman in the methods and procedures laid down by the Company for preparing Short Term Programmes.

If, at any stage during the Contract, any additional work be included in the Programme for operations not measured in the Bill of Quantities, Head Office should provide new output standards for the work involved and carry out any premeasurement of the work in order to arrive at an accurate basis for planning these extras.

SUB-CONTRACTORS AND SHORT TERM PLANNING

The various specialist work involved during the period under review will need to be carefully co-ordinated with the builders' work. It is most important that a responsible representative from the firm or firms of Sub-Contractors should be in attendance during the formation of the Short Term Programme in order that the phasing of the builders' work in conjunction with their own activities, is carefully considered.

It is useful to issue a simple Short Term Plan to the Sub-Contractors, showing the extent of their work individually plotted. The 'deadline' completions for each section, room, floor and so on, should be shown. It can then be emphasised that unless the specialist concerned keeps to his programme, the whole Contract, involving every other trade and specialist, will be disrupted.

The Sub-Contractors should also be encouraged to prepare a schedule of outstanding requirements in order that their work will be able to proceed smoothly, as and when required.

ARCHITECT AND SHORT TERM PLANNING

Depending upon the situation and the progress of the work and also the relationship which exists between the Building Contractor and the Architect, it may well be advisable to relate the Architect's Monthly Site Meeting to the Short Term Planning activities. Whatever the arrangements, every effort should be made to incorporate the developments and variations, etc., that arise from such meetings into the 'Forward View' at this stage.

REGULARITY OF SHORT PLANNING

In order that maximum benefit shall be obtained from the use of this technique, it is important for Short Term Planning to be a regular routine.

Normally, plans of this type are consecutive and the programme for a given period would be drawn up towards the end of the period covered by the previous plan.

For example, the Short Term Five Weekly Plan would be prepared at regular intervals every *three* weeks. This would mean that the overlap of two weeks would be constantly reviewed and revised to meet any changing or unforeseen circumstances.

CONCLUSION

Short Term Planning is intended to aid and assist the Site

supervision in exercising a control over the progress of work. However, a good Foreman will agree that it is far safer, as well as more business-like, to keep a chart showing what is to be carried out several weeks ahead, rather than to trust to memory. He is then released from guesswork, for not only can he tell at a glance how things are progressing, but it puts him in a very good position to answer questions at short notice.

In fact, it is true for Site Agents or General Foremen as for any other organiser, that he must give himself time to think. Systematic and properly organised Short Term Planning should provide him with both the time and the opportunity.

CHAPTER IX

WEEKLY SITE PLANNING

'. . . As soon as a man is detailed for a particular job—that is to say a duty that he has to perform . . . he may loathe the job but his reasoning mind makes him uncomfortable within himself if he neglects the job. . . .'

Rudyard Kipling.

We have considered in previous chapters how the thought processes of foresight and planning ahead weave a pattern through the management of building work. This extends from the initial thoughts at the estimating stage to overall pre-contract planning before commencing on site, and leads up to short term planning or monthly review of the progress with the Master Plan.

In order that this development can be pursued to its ultimate conclusion, proper and systematic forecasting and planning must take place regularly, on site, among the staff resident on the job who are responsible for producing the required results. In fact, the site foreman and his key personnel are a most important part of the management team in a builder's organisation. Everyone else, Estimator, Contracts Manager, Buyer, Surveyor etc. are purely a service to the man on the site. There is only one man who can organise the actual operations and keep his eye on each job to ensure that construction will proceed as specified—the site foreman. He is the man who does the honest-to-goodness practical work and enables the technician from head office to achieve his goal.

The general foreman is solely responsible for calling forward the resources at his disposal and in consequence must plan ahead in great detail to avoid unforeseen contingencies disrupting the continuity of work. Furthermore, to maintain control he must plan to maintain a proper command over the contract. This is necessary to ensure that progress is following closely on the predetermined line of events required to achieve completion of operations in accordance with estimated overall progress and costs. To do this, the site foreman must discuss the pattern of future work and make the trades foremen and specialist sub-contract foremen share in the responsibility of seeing that work is carried out in a manner to achieve the best results. This thought process and informal discussion on the site is known as Weekly Site Planning.

In order to obtain maixmum benefit from weekly planning, the general foreman should arrange regular meetings each week in order to discuss a general 'plan of campaign' for the week ahead. Each trades foreman and specialist foreman will be given the opportunity to consider a 'forward view' of the operations to be carried out during the forthcoming period. The thought process in everyone's mind at this meeting should be to anticipate problems and difficulties before they happen, in order to reduce unnecessary delays to an absolute minimum.

This opportunity to check that all requirements such as adequate labour and materials, plant, etc., will be met and proper detailed information will be available for each operation to be carried out during the period under review, often eliminates costly disruption through insufficient resources being available.

THE RELATION OF WEEKLY SITE PLANNING TO SHORT TERM PLANNING

Each month when preparing the short term plan, the site foreman and his contracts manager will usually determine 'deadline' completion dates for major or key operations. These will constantly be referred to by the site agent when assessing the amount of work that will have to be completed each week in order to maintain the rate of production required to achieve the rate of progress required.

WEEKLY SITE PLANNING MEETINGS

The keynote to successful site meetings with the builder's own trades foreman and specialist foreman is the creation of 'team spirit'. Complete co-ordination of all trades and services by collective discussion and informal debate creates the right climate for efficient and economical progress of building work. Each trades foreman must be able to identify himself as an important part of the construction team. If this attitude can be developed, pride of achievement in maintaining progress in accordance with previously arranged plans will spur the individual on to greater achievements. The whole process is a matter of the proper attitude of mind. There must be willingness at all times to co-operate and compromise with others in order to achieve the required progress.

It is the responsibility of the builder properly to train and educate his general foremen in order that they develop the important leadership qualities necessary to co-ordinate and control organised weekly domestic site planning meetings.

CHARTING THE INFORMATION FROM WEEKLY SITE PLANNING MEETINGS

Simplicity is the keynote to successful weekly planning charts. They must be readily understood by the trades foremen and the information contained on the form should only be sufficient to enable a proper control over the progress of work in each trade to be maintained.

Obviously, different types of building work call for different types of programme and therefore it is more important that the site agent and his staff shall be satisfied with the chart in use rather than head office attempting to standardise a more elaborate form of planning sheet for use on all jobs. Illustrated below are a few systems used with success on several different types of work.

THE WEEKLY SITE PLAN

The weekly site plan system is a simplified Gantt or bar line chart with the proposed week's forward load for each trade set out on a daily basis. Brief details of the gangs can be recorded and the mechanical plant that will be allocated to the gang during the week can be described. The simple method of a 'drawn arrow' across the days with descriptive notes showing what will be carried out during the period indicated will clearly illustrate to the operative the time period allowed for each operation and the deadline day for completion.

As the work proceeds actual progress can be recorded by

Figure 24.

WEEKLY SITE PLAN

CONTRACT J.S. DOBSON LIMITED. CONTRACT No. 1374. WEEK No. 17. WEEK COMMENCING 3.6.70

GANG	BUILDERS WORK	SUB CONTRT.	TRADE	PLANT	MONDAY	TUESDAY	WEDNESDAY	THURSDAY	FRIDAY	SATURDAY	SUNDAY
1ST FIXERS	6	-	CARPENTERS	HAND TOOLS	STUD PARTITIONS - (ROOM 2).				GROUNDS		
PLASTERERS No. 1.	2	-	PLASTERERS	-	M.G.	RENDER ROOM 4			SETTING	(ROOM 4)	
PLASTERERS No. 2.	-	4	PLASTERERS	-	CEILINGS - (ROOM 6).			CEILINGS	(ROOM 7)		
SPARKS	-	2	ELECTRICIAN	HAND TOOLS		FIRST FIXING			FIRST FLOOR		
PLUMBING	4	-	PLUMBERS	HAND TOOLS	H. & C. RUNS -	FIRST FLOOR.		WASTES AND C.I. WORKS G.F.			
NAVVIES	4	-	LABOURERS		CLEARING UP.	MANHOLE NO. 4.		DIGGING	OFFLOADING BKS.		
CONCRETE TEAM	5	-	LABOURERS		ROAD SOUTH SIDE 6 BAYS			CASING DRAIN RUNS - 6			

| TOTAL LABOUR | 21 | 6 | STAFF SUPERVISION | 6 | AGENT 1 | Q.S. 1 | ENGINEER 1 | G.F. 1 | FOREMEN 2 | STOREMAN - | CLERK - |

drawing another 'drawn arrow' in contrasting colour to indicate when the operation was actually carried out.

This system of weekly plan is particularly useful when the foreman requires at a glance the commitments for the week on a day-to-day basis. Each day's work is clearly set out in columns vertically across the chart. An example of the Weekly Site Plan is shown as Figure 24.

It is becoming increasingly popular among contractors to put a chart, similar to the one described, in the operatives' canteen in order to give the men an opportunity to see what the 'goal' is for the week ahead.

BLOCK PLAN LAYOUT CHART

This is the simple pictorial aid to recording decisions made at the Weekly Site Planning Meeting. The area of work being undertaken is drawn in outline to a suitable scale as a block plan. All descriptive notes and dimensions are usually excluded except in the case of stanchion bases or columns etc., when each is numbered for reference purposes. Often this sketch plan is reproduced on a silk screen duplicating machine or drawn on tracing paper and duplicated in a similar manner to architectural details. When the block plan will be used continuously as in the case of a multi-storey building, a purpose-made rubber stamp can be prepared for a few pounds and will undoubtedly prove to be a sound investment.

The method of recording progress with this system is to superimpose the planned weekly work load for each trade in coloured crayon over the block plan, in different colours.

Each trades foremen will have a copy and will be able to see readily the amount of production required in order to enable all other trades to make their contribution to the overall progress.

The advantage in this system of recording information is, obviously, its simplicity.

All trades foremen are used to interpreting working drawings in order to discover the method of construction required. This system also enables the foreman to refer to a drawing in order to see at a glance how long he has got to carry out the work. This system of weekly planning is popular amongst house builders and is often used as a flow diagram showing sequence

Figure 25.

WEEKLY SITE PLAN

CONTRACT					CONTRACT No.	WEEK No.		WEEK COMMENCING	

GANG	BUILDERS WORK	SUB CONTRTR	TRADE	PLANT	PLAN	ACTUAL	REMARKS

NOTE.

UPON THIS TYPE OF CHART THE WORK
LOAD FOR EACH TRADE DURING THE
WEEK CAN BE PLOTTED IN CONTRAST-
ING COLOURS.

THIS METHOD OF WEEKLY PLANNING
IS VALUABLE DURING THE CONSTRUCT-
ION OF THE FOUNDATIONS AND SUPER
STRUCTURE WORK.

	STAFF SUPERVISION	AGENT	Q.S.	ENG.	G.F.	FOREM.	CLERK	STOREM.	WATCHM.
TOTAL LABOUR									

of hand-overs and direction of the progress of work across the estate.

Actual progress can be recorded in a contrasting colour on the plan. This, together with the date stamp will provide a most useful record for the Builder for reference when settling claims at a later stage. A copy of the Block Plan Layout Chart is shown used in conjunction with a Weekly Site Plan in Figure 25.

ISOMETRIC PLANNING CHARTS

The Isometric Planning Chart is a development of the previous system of Block Plan Layout Charts. This system enables the foreman to relate work being undertaken on different floor levels. The isometric drawing shows an outline for a multi-storey building and is shown as Figure 26.

This is clearly a visual aid to Weekly Site Planning. It will be seen that the week's work content for each trade can be shown as described for the Block Plan Chart and, in addition the flow of work can be illustrated between various floors.

The system is particularly useful during the finishing stages in multi-storey work. Provision can be made on the chart for outstanding requirements for the week and, particularly during finishing work, for a note about quality points to be carefully considered during the week.

INFORMATION FROM WEEKLY PLANNING CHARTS

If the Weekly Site Planning Meeting has been properly conducted and the plan prepared efficiently, information will be available about the week's work ahead for everyone concerned.

1. *Sequence of Operations*

   A complete and accurate sequence of operations for each trade for the forthcoming week will be established. This work will be shown properly co-ordinated with other trades to produce a comprehensive programme.

2. *Operatives*

   The number of people required to undertake the work in each trade will be apparent. Arrangements can be made in advance to increase or decrease the labour force as necessary.

3. *Plant*

   Mechanical equipment not yet available on the site can be

Figure 26.

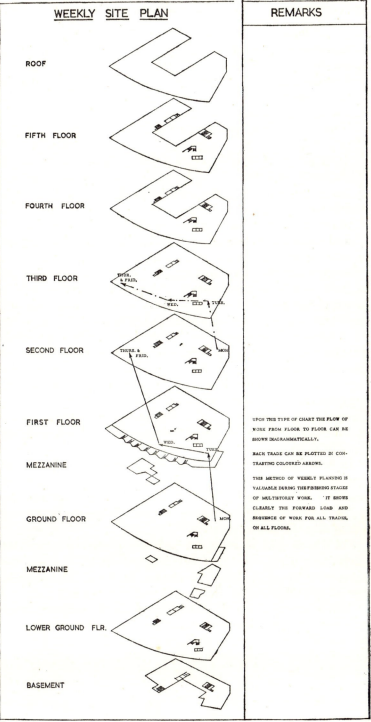

| WEEKLY SITE PLAN | REMARKS |

ROOF

FIFTH FLOOR

FOURTH FLOOR

THIRD FLOOR

THUR. & FRID.

WED.   TUES.

SECOND FLOOR

THURS. & FRID.

MON.

FIRST FLOOR

WED.   TUES.

MEZZANINE

GROUND FLOOR

MON.

MEZZANINE

LOWER GROUND FLR.

BASEMENT

UPON THIS TYPE OF CHART THE FLOW OF
WORK FROM FLOOR TO FLOOR CAN BE
SHOWN DIAGRAMMATICALLY.

EACH TRADE CAN BE PLOTTED IN CON-
TRASTING COLOURED ARROWS.

THIS METHOD OF WEEKLY PLANNING IS
VALUABLE DURING THE FINISHING STAGES
OF MULTI-STOREY WORK. IT SHOWS
CLEARLY THE FORWARD LOAD AND
SEQUENCE OF WORK FOR ALL TRADES,
ON ALL FLOORS.

H

called forward and a firm date when it will be required announced with confidence.

4. *Material*

Suppliers can be notified a week ahead and consignments of material agreed and suitable times for delivery arranged.

5. *Outstanding Information or Queries*

Every technical detail should be to hand at this stage of the job. Any outstanding information or query must be settled at once. This system of weekly planning usually provides the builder with his last chance to call forward his requirements, before outstanding details cause delay.

6. *Methods*

All methods such as material handling etc. must be agreed and finalised by this time. Everyone concerned should be fully aware *how* the job will be done.

No Agent or General Foreman, however experienced he may be, will be able to rely upon his memory for what he will do 'on site next week' for every operation in every trade. Nor will he be able to ensure that all requirements for the following week are available for all operations. If a properly conducted Weekly Site Planning Meeting is arranged, and a complete 'plan of campaign' agreed with trade foremen or sub-contract foremen, together with a full and detailed 'list of outstanding requirements' the work on site the following week will flow smoothly. The Agent or General Foreman will be able to devote more time to supervision instead of spending the week jumping from crisis to crisis.

In order that the builder will be able to obtain full benefit from weekly site planning, the foreman must be able to control not only the amount of work to be undertaken but also the limits of expenditure, to enable the week's work to be carried out within the estimated costs.

For a plan to be of any material value, it must be effectively controlled. This can be achieved by checking the physical progress of the plan and by measuring the financial value and cost of the plan against the actual results achieved.

This is a form of Financial Budgeting and is complementary to Weekly Site Planning which will be described in Chapter X.

FINANCIAL BUDGETS

'The lack of money is the root of all evil.'

Mark Twain.

The traditional methods of accounting commonly used by builders, while always necessary in order to determine whether or not the firm is trading at a profit, seldom reveal sufficient information in time to enable the management to take action to control costs.

Annual or quarterly accounts, or the review of the financial position of contracts periodically and then on completion, are really statements of historical interest. It has even been said that these costs are only of value to the Bank Manager. The Contracts Manager receives them and is often made aware of the situation only 'after the horse has long since bolted'.

In order to adopt a system that will achieve control, the information must be available before the work takes place. In other words, the builder must plan in advance what his costs ought to be. Many builders will say that it is impossible to forecast accurately what costs are going to be on a particular contract. This may be true, but on the other hand if an estimator can calculate a firm quotation for the work, then the builder should be able to determine what his costs ought to be in order that the work will show a profitable return for the capital outlay in undertaking the work. This is the principle of budgeting—or planned expenditure, A translation of a programme into a financial forecast.

This principle of managerial control must emanate from the top. At Board level, an accurate budget of expenditure will reduce the risk of over-trading—for example, undertaking work of such value that it would be in excess of the limits of turnover permissible with the company's working capital.

TURNOVER BUDGETS

Let us then consider an example of the process of forecasting by the Board of Directors of a building company. Their task is to examine the present trends of the market, by analysis of information made available through research or with the assistance of statistics published from time to time. They will attempt to forecast future demands that will be made for building work over a defined period of time; i.e. five years.

We will take the purely hypothetical case of the builder who has the following information available.

In 1966 it is calculated that demand for private enterprise houses in an area within 24 km radius of his headquarters will have risen from £10 million per annum in 1961 to £11½ million per annum in 1966. At the end of his trading year in 1961 the turnover of the company was £500 000, this represents 5% of the market turnover in the area. Bearing in mind the anticipated increase in the costs of labour and materials in addition to the potential increase in market value of the houses he builds over the next 5 years, he forecasts that he will expect to undertake 7% of the market turnover in 1966. 7% of an anticipated market value of £11½ million = £800 000 turnover for the builder. The effective increase annually is shown in this table:

Year	1961	1962	1963	1964	1965
Turnover	500 000	575 000	650 000	725 000	800 000
Increase		75 000	75 000	75 000	75 000

Increase % annually 15% for 4 years shows increase of 60% total.

From this simple example, it will be seen that turnover, profits and costs can be calculated; and also the necessary working capital required in order to trade effectively, thereby providing the Board of Directors with a turnover budget for 5 years ahead.

Always an important consideration to the Builder is the financial demand that a contract will make on his capital. This is particularly true if the job is large and extends over a considerable period. In order to forecast accurately the amount of money required, the contractor must make an attempt to calculate the rate of expenditure and compare this with anticipated interim payments. Obviously the preparation of such a financial schedule can only be approximate, as a high degree of accuracy is impossible due to unforeseen contingencies creating large variations in the work value.

However, a simple system of calculations will provide a useful guide to the builder in attempting to determine his overall financial commitment for a contract. We will now consider the case of a hypothetical contract to be undertaken over a period of, say, twenty-four months. The first consideration must be the determination of the total costs on a monthly basis. To start with the initial expenses such as deployment of site, installation of office accommodation, etc. must be determined. Expenses such as site oncosts and overheads must be spread out uniformly over the total contract period. Then the estimated cost of each operation is distributed evenly over the time period allowed in the overall programme of work. For simplicity all the costs have been reduced to the nearest ten pounds.

INITIAL COSTS

A. Installation of site equipment $= £3\,000$
    Layout of materials.

B. Overheads, Insurance, Supervision, etc.

$$\frac{£13\,381 + £5\,680 + £10\,173}{24} = £1\,200 \text{ per month}$$

C. Building Operations
1. Site Stripping
    £977 for 2 months $=$ £480 per month
2. Excavation, foundations etc.
    £2 984 for $4\frac{1}{2}$ months $=$ £670 per month
3. Concrete work
    £19 545 for 12 months $= £1\,630$ per month
4. Carpentry Work
    £3 276 for $13\frac{1}{2}$ months $=$ £240 per month
5. Joinery and Fittings
6. £7 427 for 6 months $= £1\,240$ per month
    Steelwork, etc.
    £58 277 for 12 months $= £4\,860$ per month

7. Finishings to walls and floors
   £1 438 for 2½ months   =   £570 per month
8. Decorations, etc.
   £4 803 for 5 months   =   £960 per month

The next stage in the calculations is to compute the forecast of income to be derived from interim stage payments. This will represent the value of monthly certificates less payments to sub-contractors. The object is to find the value of the work for each operation, and uniformly distribute this over the time period allowed in the Overall Programme for the work, in a similar manner as when calculating the costs per month.

The monthly payments that the general contractor must make to his sub-contractors is computed by dividing the value of the sublet work by the time allowed to execute the work as follows:

1. Stonework to Elevations
   £47 200 for 13½ months   = £3 500 per month
2. Metal Windows
   £571 for 2½ months   =   £230 per month
3. Suspended Ceilings
   £2 228 for 3 months   =   £740 per month
4. Decorative Panels
   £498 for 4 months   =   £120 per month
5. Plastering
   £21 060 for 8 months   = £2 630 per month
6. Tiling
   £1 933 for 5 months   =   £390 per month
7. Roofing
   £18 277 for 8 months   = £2 280 per month
8. Flooring
   £2 262 for 5½ months   =   £410 per month
9. Acoustic Tiling
   £2 801 for 5 months   =   £560 per month
10. Painter
    £8 340 for 7½ months   = £1 110 per month
11. Glazing
    £6 231 for 7 months   =   £890 per month
12. Terrazzo
    £8 410 for 6 months   = £1 400 per month

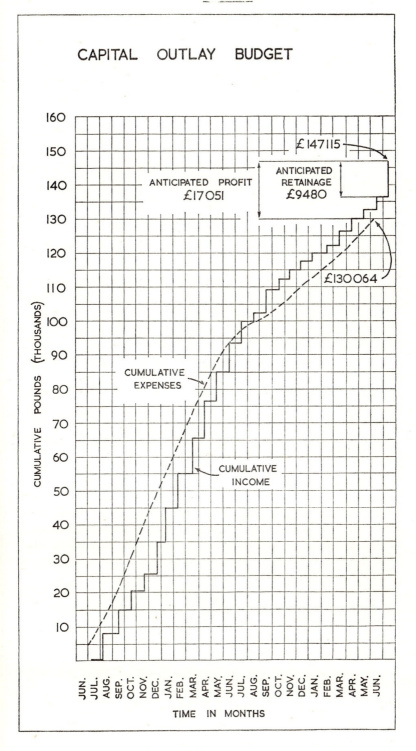

Figure 27.

13. Heating Installation
    £51 200 for 24 months   = £2 130 per month
14. Electrical Installation
    £33 876 for 24 months   = £1 410 per month
15. Painting Externally
    £7 933 for 3½ months   = £2 260 per month

The graph shown as Figure 27 shows the cumulative income budget representing interim payments by the client based upon the receipt of the cheque on the 10th of each month.

In the contract, a retention of 10% is a condition and the limit of retention is when 50% of the work is completed. The graph shows the total value of income less 10% for the first eleven months of the contract.

At this stage, it will be seen that the value of retention is £9 480 for the remaining thirteen months and the interim payments are reduced by this figure as no further retention is being made.

From this information plotted on the graph, the builder is able to see how much money will be required to carry on the job and when the demand will be heaviest.

To produce a similar type of financial budget for all work to be undertaken, the builder can prepare an overall cash forecast of the capital requirements of the company. The amount and duration of short-term cash loans can be determined with reasonable confidence.

CONTRACT EXPENDITURE BUDGET

By calculating the estimated site costs before work commences, the contract management is able to exercise a control over the work and to moderate the expenditure of labour etc. in order to keep the actual site cost within the limits of the estimate for the work.

To prepare a planned expenditure budget for the site which is geared to the programme of work and the estimate, will enable the builder to locate excessive costs or actual losses as the work is in progress. Action can then be taken *at once* to avoid or reduce further loss.

In contract budgeting, the Overall Master Plan is expressed

Figure 28.

FINANCIAL BUDGET

CONTRACT J.S. DOBSON.   CONTRACT No. 1374.   OPERATION DRAIN LAYING (MANHOLE 1 - 2).

No.	DESCRIPTION	VALUE £	WEEK NO. 1.	2.	3.	4.	5.	6.	7.	REMARKS - SPECIAL CONDITIONS
1	SETTING OUT.	£ 5.00	£ 5.00							
2	EXCAVATE TRENCH.	£ 75.00	£ 25.00	£ 25.00	£ 25.00					
3	LAY PIPES AND LEVEL.	£ 25.00		£ 12.50	£ 12.50					
4	TEST PIPES.	£ 5.00				£ 5.00				
5	ENCASE DRAIN WITH CONCRETE.	£ 60.00				£ 20.00	£ 20.00	£ 20.00		
6	BACKFILL TRENCH.	£ 30.00						£ 15.00	£ 15.00	
	TOTAL VALUE	£ 200.00	£ 30.00	£ 37.50	£ 37.50	£ 25.00	£ 20.00	£ 35.00	£ 15.00	

SHOWING (a) TOTAL VALUE FOR EACH OPERATION
(b) CALCULATED VALUE OF WORK COMPLETED EACH WEEK

in money value obtained from the priced Bill of Quantities or Estimate.

TOTAL VALUE OF WORK

By allocating total values from items in the Bill of Quantities to each operation in the programme and allocating a proportion of the cost of the operation to each week during which it is planned to be executed, it is possible to determine the total value of work forecast for completion weekly. This is illustrated by a simple example, shown as Figure 28 for digging drain trenches.

This system of budgeting can help the builder in preparing applications for interim payments; furthermore it will encourage the builder's surveyor to 'value-up' his application instead of the common practice of keeping some 'up one's sleeve' for the final valuation. To obtain *maximum* interim payments on all contracts will greatly ease the burden and tend to reduce industrial and bank loans to a minimum.

SITE LABOUR COSTS

To extract only the estimated site labour costs from the estimate, and allocate these on a weekly basis in conformity with the programme of work, will provide a useful check upon the actual weekly site wage bill.

Any drastic deviation from the planned or budgeted labour costs will bring about a critical investigation by the management into the progress and efficiency of the site and the possible accuracy of the estimating output values. To expect the site foreman to control work at a profit without providing him with the estimated limits of what his costs must be is like giving him a watch without any hands and expecting him to tell the time.

PLANT HIRE COSTS

Mechanical equipment is an expensive part of construction costs and must be used to maximum capacity with the builder benefitting from the greatest possible production and the minimum of cost. To ensure this the estimated plant hire charges for each operation on the programme must be determined. Each week the site must be informed what estimated hire cost of equipment is allowed for the programme of work

Figure 29.

DAILY   OPERATIONAL   COST   SHEET

CONTRACT   J.S. DOBSON LIMITED.   CONTRACT Nº.   1374.

DAY & DATE   20TH NOVEMBER, 1970

GANG   NAVVIES.

NUMBER   6

CHECK Nº.	NAME	TRADE	OPTN Nº. 2	3	4	5			TOTAL HOURS
			HOURS PER OPERATION						
421	THORNE P.	L	4	$4\frac{1}{2}$					$8\frac{1}{2}$
422	MOORE F.	L	2	$6\frac{1}{2}$					$8\frac{1}{2}$
423	VARLEY D.	L	$6\frac{1}{2}$	1	1				$8\frac{1}{2}$
424	MORCOM D.	L	4	4	$\frac{1}{2}$				$8\frac{1}{2}$
425	DAVIS E.	L	2	6	$\frac{1}{2}$				$8\frac{1}{2}$
426	HOLLOWAY P.	L	2	2	2	$2\frac{1}{2}$			$8\frac{1}{2}$
TASKS COMPLETE INDICATE BY C									

REMARKS

WEATHER FINE - NO STOPPAGES.

SIGNATURE   R. M. R.

Figure 30.

WEEKLY OPERATIONAL COST SHEET

CONTRACT: J.S. DOBSON. CONTRACT No. 1374. TRADE: LABOURERS. W/E DATE: 10.4.70.

NAME	DAY OPTN.	MON. 1 2 3 4 5 6	TOTAL	TUES. 1 2 3 4 5 6	TOTAL	WED. 1 2 3 4 5 6	TOTAL	THUR. 1 2 3 4 5 6	TOTAL	FRIDAY. 1 2 3 4 5 6	TOTAL	SAT. 1 2 3 4 5 6	TOTAL	TOTAL HOURS
P.J. WADE.		4 2 2	8	8	8	4 4	8	2 2 2	8	8 4	8		4	44
M. MYRES.		8	8	8 4 4	8	2 6	8	8	8	8 2 2	8		4	44
E. HUBBALL.		3 5	8	3 5	8	3 5	8	8	8	2 6 3 1	8		4	44
J. STANTON.		8	8	8	8	2 6	8	2 4 2	8	2 2 2 1 3	8		4	44
E. HARTLAND.														

OPERATION

1.	OFFLOADING BRICKS.
2.	EXCAVATING MANHOLE 4.
3.	ASSISTING CARPENTER TO ERECT COLUMN SHUTTERS.
4.	HANDLING STEEL REINFORCEMENT.
5.	
6.	

planned. This will often encourage the use of the builder's own plant instead of employing equipment at expensive external hire rates.

The economical use of all builder's plant is the keynote to efficiency on modern building sites, particularly upon multi-storey work where expensive cranes and hoisting equipment is employed.

The budget of plant costs against each planned operation will ensure maximum productivity from the equipment. In addition it will be simple to diagnose the uneconomical use of plant which exceeds the estimated plant costs in the estimate.

RECORDING ACTUAL COSTS ON SITE

In order to ensure that a proper comparison is obtained between actual costs and budgeted expenditure, a reliable system of site records is essential. Much will depend on the integrity of the site staff to allocate accurate costs to the various operations. No benefit to the company can be gained from fraudulent records or the 'cutting and loading' method of recording actual site costs.

On the other hand, top management must realise that if the site forman fears criticism from submitting records which show a loss, then he is likely to present attractive but inaccurate pictures of the progress of work in order to avoid personal consequences for producing 'unpopular' information.

DAILY OPERATION COST SHEET

This is a system of recording on a daily basis, actual costs of operations. Usually the trades foremen carry a small book with them to record the hours spent and the operations undertaken by the men under their control. The general foreman will normally check these records daily against the wage sheet hours to ensure that all the time is allocated. An example of the Daily Operation Cost Sheet is shown as Figure 29.

WEEKLY OPERATION COST SHEET

This is similar to the Daily Operation Cost Sheet but pro-vides for a full working week of actual costs to be summarised on one sheet. An example of the Weekly Operation Cost Sheet is shown as Figure 30.

Figure 31.

OPERATIONAL COST SHEET

CONTRACT	J.S. DOBSON LTD.	CONTRACT Nº 1374.	RECORDED BY K. HILL.	DATE 2.6.70

BRIEF DESCRIPTION OF OPERATION	LOCATION	WORKING CONDITIONS ON SITE	DIMENSIONS & DETAILS SIZES OF OPERATION	SPECIAL CONDITIONS & REMARKS
EXCAVATING BASEMENT AREA.	SOUTH EAST SIDE.	DRY FINE.	SIZES OF OPERATION : 56m x 18m 7.5m	SANDY GRAVEL. FIRM.

NAME	TRADE	W/E M	T	W	T	F	S	S	W/E M	T	W	T	F	S	S	W/E M	T	W	T	F	S	S	W/E M	T	W	T	F	S	S	OFFICE USE ONLY
MOUNTFORD S.	L	8	8	8	8	8	8	8																							88
SHROPSALL R.	L	8	8	8	8	8	8	8																							80
DENNIS P.	L	8	8	8	8	8	8	8																							80
HIPKINS S.	L				8	8	8	8																							40
																															288 hours @£0.45 = £129.60

TYPE OF PLANT		W/E M	T	W	T	F	S	S	W/E M	T	W	T	F	S	S	W/E M	T	W	T	F	S	S	W/E M	T	W	T	F	S	S	
JCB 6.C.	W	6	7	8	4	6	3	7	7	7	6	4	7	2																	67 @ £2.50 = £167.50
	I	2	1	-	4	2	1	1	1	2	4	1	1	2																	21 @ £1.00 = £21.00
	B																														88 hrs. £188.50
10 R.B. DRAGLINE.	W							6	8	7	7	6	2																		36 @ £2.00 = £72.00
	I	2	-	1	1	2	2																								8 @ £1.00 = £8.00
	B																														44 hrs. £80.00
	W																														
	I																														TOTAL PLANT £268.50
	B																														

Figure 32.

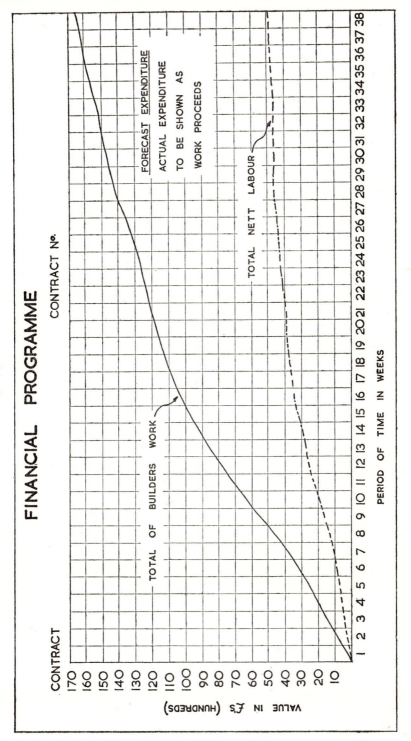

OPERATION COST SHEET

This type of pro-forma is popular for civil engineering work but can be equally effective in recording information to be compared later with planned budgets of expenditure. Here labour and plant costs are recorded together, and provision is made to calculate the actual unit cost of labour and material for the operation involved. See Figure 31.

GRAPHICAL COMPARISON CHARTS

Some builders find that a simple graph showing the planned budget for labour, plant or total value of work is the best visual aid to cost control. The graph can be drawn up to show on a weekly basis the budgeted cumulative costs. Actual site costs plotted on the graph each week in a contrasting colour show at a glance the financial position of the contract and enable prompt action to be taken when a deviation from the budget occurs. See Figure 32.

CONCLUSION

The results produced by this control will show the profitability of site operations as a whole. Should the results show an unprofitable situation, investigation will be necessary into the details of the work planned, or produced, so that the cause can be found.

This control will not provide a means of measuring the effectiveness of individual site operations or of checking the validity of budgeted costs in tender prices. The potential economies through more efficient performance of individual operations are considerable, and will only become available through the application of Work Study.

CHAPTER XI

INCENTIVES

'. . . The great incentive, fear of unemployment in industry, has gone, let us hope never to return . . .'

Frank Russon.

In Chapter IX we dealt with the technique of Weekly Site Planning and considered the process of systematic management by the site staff. For full benefit to be obtained from successful leadership by the foreman, every opportunity must be taken to encourage the operative to increase productivity. A properly integrated incentive scheme could play a most important part in establishing more effective co-ordination and co-operation between operative and the management.

It is important, if a programme of work is to be achieved, that the operative must be encouraged to approach his work with enthusiasm and with drive to complete his part efficiently and with minimum delay.

In the industry as a whole, incentives are by no means established and unfortunately it is not possible to deal with the whole subject of incentives in the Building Industry in a single chapter.

We will consider, however, several possible reasons why systems of incentives already tried have failed and, briefly, the main points covered by Special Report No. 28 of the National Building Studies 'Incentives in the Building Industry'.

RESPONSIBILITY FOR INCENTIVE SCHEMES

Where companies have introduced systems of bonus and incentive payments, it has usually been regarded as an additional responsibility for the surveying staff—a chore to be undertaken between valuations and often only carried out when availability of labour was difficult. It is important that incentives are regarded as a central process in the organisation of the

I

firm. Staff must be properly trained and given the necessary status to ensure that the function is carried out effectively.

PURPOSE OF INCENTIVE PAYMENTS

Incentive payments to operatives are often regarded as a mere 'bribe' to retain labour. Many schemes now in operation would be abandoned if labour became more plentiful; presumably on the assumption that fear of dismissal would be sufficient incentive to work hard. This view is most unfortunate, because enormous advantages are to be gained when payment by results is used as a tool of management to help the organisation to achieve pre-determined progress within specified costs.

ORIGIN OF INFORMATION USED FOR OUTPUT STANDARDS

Possibly the greatest pitfall for all incentive schemes is lack of accurate and reliable output standards.

Considerable thought has been given by many authorities on the subject of design of a working procedure for incentive schemes. No system will produce the desired results unless the incentive placed before the operative is based upon an output that can be expected from an average man, under average conditions, working at a normal rate.

Most output values are based upon opinions. Some may be related to the estimator's unit labour rate in the estimate for the job. This is often regarded as undesirable practice.

The only successful systems of setting realistic output targets is by time measurement of the operation concerned or actual unit costing from work of an identical nature.

Space does not permit discussion of the value of Work Study in this field of incentives, suffice it to say that unless bonus targets are based upon some accurate form of Work Measurement, any system of incentives introduced on a building site is likely to fail.

A slogan that should be adopted is 'Let's get the target right —then the increased output will be automatic'.

PAYMENT OF SAVINGS

It is the practice in some firms to pay a percentage of the savings as bonus. This in the writer's opinion is psychologically wrong. If the man on the job has got to share his bite of the carrot with his boss he is going to wonder whether it is worth the effort in the first place. If the Contractor feels that he must retain a proportion of the saving to cover the costs of the administration of the scheme, then he should adjust his *target* to reduce the margin of savings shown. This method is not wrong, where targets are guesses, since it limits the worse effects of over payment.

It is often important that 100% of any savings should be paid to the operative. If this practice is followed then increased productivity should be automatic.

Mr. Frank Russon, in his book *Bonusing for Builders*, published in 1950, urged that an exchange of knowledge and experience between firms of all sizes should be adopted with the ultimate aim of standardising administrative methods.

It was not until 1958 that H.M. Stationery Office published 'National Building Studies Special Report No. 28'. In it, the essential requirements of a good incentive scheme are emphasised. 'From this study two main aspects of incentive schemes can be distinguished: the design of the scheme itself, including methods of recording and site operation, and the relation of incentives to the firm's organisation as a whole, including the pattern of responsibility for decisions within the scheme as related to the firm's own organisation structure.

'On the first of these aspects, design of the scheme, the principles brought out by the study are:

1. Operation targets, coupled with a recording system giving operation costs, should be used whenever possible.

   Operations should be, as far as possible:

   a. visual stages of work;

   b. of about one week's duration

   c. continuous jobs with no hold-ups for other trades.

2. The systematic use of cost information from all sites and close consultation with site staff contribute to maintaining accurate and well-balanced targets.

3. The incentive effect of the scheme can be increased to the maximum by:

   a. simplifying the principles of the scheme and the recording system and explaining these adequately to the operatives so that they can calculate their own bonus;

   b. employing small gangs as bonus units;

131

    c. adhering to the payment of the actual amount earned under the target by eliminating extra payments and by not putting an upper limit on bonus;

    d. extending the coverage of the scheme as far as possible over work on the site;

    e. making bonus payments weekly and as soon as possible after the completion of the operation.

4. Good labour relations on the site can be encouraged by:

    a. presenting the targets to the operatives for agreement before the work starts;

    b. establishing a recognised channel for complaints;

    c. maintaining a simple and direct scheme.

5. It is necessary to safeguard the quality of work by:

    a. making site staff independent of the production bonus;

    b. improving site supervision of quality.

6. The scheme should be designed to suit individual needs and local conditions. This calls for flexibility in:

    a. The percentage of saving paid as bonus. It is preferable to divide the saving between employer and operatives, but local conditions on particular sites may lead to better results with a hundred per cent payment to the operatives;

    b. Distribution of bonus. The operatives on each site should decide the method of sharing within the bonus group or gang.

    c. Target adjustment. Targets should be eased if adequate reason is established on a particular site;

    d. Organisational methods. A review of the whole scheme should be made regularly, perhaps once a year, to ensure efficient operation;

    e. Treatment of disputes. These require quick decisions at high level, reached in close consultation with site staff.

'The reason why these principles are not widely applied in practice can often be traced to organisational factors. In particular, consideration of the flexibility required in operating the scheme reveals the need for well-defined channels or communication both within the organisation of the incentive scheme and also in its relation to the other functions of management.

'The study has emphasised the separation that often exists in building firms between tendering, planning and production, and has indicated the important part incentives could play in establishing a more effective co-ordination. Thus it is important that incentives be given a central position in the organisation of the firm and that staff of adequate status and ability be employed in the bonus system. This is particularly important at the planning stage on each contract, when decisions on the form of the scheme to be employed should be brought into line with the site management and planning processes so that the incentive scheme can be operated with a realistic understanding of the existing and the possible site problems. Information used for the bonus scheme can also provide the basis for programming at the planning stage, and, throughout the contract, gives a systematic control of cost at operation level. The incentive scheme should also maintain a long-term connection with estimating in order that the cost information used by the estimator should be as realistic as possible and that tender prices should be related to the general expectation of labour costs.'

CHAPTER XII

CONCLUSIONS—MANAGERIAL LEADERSHIP

No amount of attendance at lectures and courses will give a man those personal qualities essential in the make-up of an efficient Manager in the Building Industry, nor can they be obtained from books. It is not a question of how much we may hear or read but of how much knowledge drawn from others' experience we can turn over in our minds, convert into power and apply to our own circumstances.

The vital qualities of personality and leadership cannot be developed without outside help. Initiative and the power to delegate and get things done cannot be taught or imparted, but these essential inherent qualities for the management of men can be self-developed.

This book has dealt with technical procedures intended to add to the building executive's knowledge, to assist him to handle his responsibilities more intelligently. He is asked to consider himself and his job and appraise himself and his efficiency as a *leader*.

There are few subjects in the field of management that have been exploited more than the topic of leadership. This last chapter is an opportunity to dwell briefly upon the great value of this personal quality which is essential if the principles considered in the preceding chapters are to be applied to advantage.

Poor leadership is not uncommon in building firms today. Many contractors have failed because of the poor standard of leadership. In some cases it may be traced to the unwillingness of the management to delegate authority to other persons or to trust them with responsibility. Poor leadership may be the result of jealousy of the capacities of others, or of reluctance to train junior members of the organisation to occupy executive positions.

This lack of leadership is prevalent in those businesses where the owner himself has control and undertakes all the major functions of the business. The activities of such a business are limited by the calibre of leadership and creative ability of one person.

Let us consider the attributes of creative executive leadership.

Good executive leadership is the ability to create ideas and instil into every member of the organisation a sense of confidence, loyalty, willingness, satisfaction and co-operation. The final proof of a good leader is his ability to handle people properly in all types of situations. He must be a thinker and able to motivate others to act!

We have dealt at some length with the qualities one consciously or unconsciously looks for in a person entrusted with control of the working lives of others, and we shall consider later another essential requisite—the ability to co-operate. At the moment we wish to stress the necessity of facing up to the job with sincerity and moral courage, the determination to be just and honest in all dealings. It is difficult to be fair and impartial at all times. Yet the increasing standards required of men in supervisory positions demand that subordinates should know, and not merely hope, that whatever else happens, they will get a 'square deal'. If a Manager knows he is sensitive to criticism or even comment; if he is inclined to form personal likes or dislikes or to be swayed by anything but facts and circumstances, he must be on his guard against decisions influenced by these characteristics. Irritability and emotion are crucial weaknesses, whereas stability and a reputation for being fair-minded inspire confidence and earn respect. Before making a decision, he must put himself in the other man's place—whether it be a subordinate or a superior—and ask himself what he would do then. It will sometimes be necessary for him to criticise and even reprove, but it can be done in such a manner as not to antagonise an ordinary person and not to leave a feeling of unfairness. A manager must discriminate between the important and the less important and not have pet bugbears. The men must be able to depend on him. Reliability applies not only to the quality of sticking to a job and seeing it through to a successful conclusion, but to keeping his word and his promises—his word is his bond. This does not mean that he must be stubborn and inflexible, but that he can be trusted and relied on.

He must not follow a system blindly, but he will not say one thing today and another tomorrow. Frequent changes of opinion do not indicate progressiveness but that ideas are not clear or stable. A good manager will give way honourably on the score of reason.

If he is unruffled when things go wrong, not influenced by opinions but only by facts, he will be able to stick fairly to his decisions. If he puts his heart into his job, showing that he is willing to do his best, trying at all times to see his subordinates' difficulties and displaying an attitude of helpfulness, he will find that ordinary responsible workpeople mould themselves to his example. If he shows himself loyal to the firm and has the interests of his 'job' at heart, he will find his subordinates loyal to them also.

The possession of authority is a hard and searching test. It strengthens strong wholesome characters; it puffs up or breaks the weak. How does a Manager respond to being in a position of authority and responsibility? Does it make him feel what a fine fellow he is and thirsty for adulation, or does it impress him as a greater opportunity for being of service to others, service to his subordinates, service to his firm and in a wider sense, service to the community?

With the increasing demands made on men in senior executive positions in the building industry, a broader outlook and administrative capabilities, not mere technical knowledge, are becoming essential.

The executive, or student, must keep himself up to date on what the managements of other firms and other industries are doing.

For this reason the Bibliography includes periodicals which men in positions of responsibility should endeavour to see regularly.

Some of the books mentioned are fairly advanced and do not specifically deal with the building industry, but there are none which will not repay the ambitious and energetic executive or student.

BIBLIOGRAPHY

This list is for those wanting further information on the subjects touched on in this book. It includes, but is not restricted to, acknowledgements of publications quoted in the body of the book.

JOURNALS
The Manager. 80 Fetter Lane, London E.C.4.
Business. 109 Waterloo Road, London S.W.1.

HISTORICAL DEVELOPMENT
The Making of Scientific Management. L. Urwick and E. F. L. Breck (Pitman).

MANAGEMENT IN PRINCIPLE
Golden Book of Management. L. Urwick (Newman & Neane).
Scientific Management. F. W. Taylor (N. Y. Harpur).
The Principles and Practice of Management. E. F. L. Breck (Longmans).
The Writings of the Gilbreths. W. R. Spriegal and C. E. Myers (U.S.A.: Irwin).
Fundamentals of Professional Management. J. G. Glover (U.S.A.: Boardman).
Elements of Administration. L. Urwick (Pitman).
Management, its Nature and Significance. E. F. L. Breck (Pitman).

MANAGEMENT IN PRACTICE
Management in Practice. P. F. Drucker (Heinemann).
Site Records for Builders (H.M.S.O.).
Contractors' Plant, its Organisation, Operations and Maintenance. H. O. Barratt (Pitman).
Outline of Work Study Part I and II. (B.I.M.).
Incentives in the Building Industry Report No. 28. H.M.S.O.
Bonusing for Builders and Allied Trades. F. Russon (Tiptaft).
Civil Engineering Contracts Organisation. John C. Maxwell-Cooke. Cleaver-Hume Press Ltd. London W.8.

PERSONNEL MANAGEMENT
Human Factor in Management. S. D. Hoslett (Macdonald & Evans).
Personnel Management in Perspective. L. Urwick (Institute of Personnel Administration).
Interviewing for Selection of Staff. Anstey and Mercer (Allen & Unwin).

FINANCE

General Financial Knowledge. E. M. Taylor and C. L. Lawton.
Introduction to Budgetary Control. (Institute of Cost and Works
Accounts).
Budgetary Control and Standard Costs. J. A. Scott (Pitman).
Standard Costs. H. E. Kearsey (Pitman).

MISCELLANEOUS

Business Enterprise. R. S. Edwards and H. Townsend (Macmillan).
Motion and Time Study. R. M. Barnes (Wiley).
A Fair Day's Pay. J. J. Gracie (Pitman).
Modern Foremanship. T. H. Burnham (Pitman).
Work Study in the Office. H. P. Cemach (Office Magazine).
Programming and Progressing. (H.M.S.O.).
Organising for Production. B. H. Dyson (Industrial Administration
Group).
Management Survey. F. Hooper (Pelican).
"Housebuilding with a Tower Crane." The Municipal Journal,
Sept. 11, 1953.
"Building Progress Schedules and Programmes." R. H. James,
The Architects Journal, Feb. 1, 1951.
"The Use of Mobile Cranes for Housebuilding." R. H. James,
The Architects Journal, June 21, 1951.
"Work Study in the Building Trade." Geoffrey McLean, Time
and Motion Study, Oct. 1953; Nov. 1953.
"Efficiency in Building." O. J. Masterman, The Architects Journal,
Oct. 20, 1956.
"What an Architect Expects from a Contractor." Sir Thos.
Bennett, The Builder, June 29, 1956.
"Site Organisation and Design at Trinity Road Flats, Wands-
worth." (Reprints for M.O.W. Information Office.) The
Architects Journal, Feb. 3, 1955.
"The Programming of House Building." Building Research Digest,
No. 91, Aug. 1956.
"The Architect and Contract Planning." A review of Building
Research Digest, No. 91, The Architects Journal, May 16, 1957.
Ministry of Works Publications:
  (a) Programme and Progress. H.M.S.O., 1944.
  (b) Production in Building and C.E. Analysis of Manhours.
      H.M.S.O., 1946.
  (c) Site Records for Builders—I. H.M.S.O., 1952. (House
      Programming.)

INDEX

141